# Dale Carlson's
## AWARD WINNING NONFICTION AND FICTION
### For Teens & Young Adults

**3 ALA Notable Books • Christopher Award**
**3 New York Public Library Best Books for Teens**
**VOYA Honor Book • YALSA Nominated Quick Picks for Teens**
**2 *ForeWord Magazine* Best Book of the Year Awards**
**International Book of the Month Club Selection**

PUBLISHERS WEEKLY calls Carlson's ***Girls Are Equal Too: The Teenage Girl's How-to-Survive Book***, "Spirited, chatty, polemical…A practical focus on psychological survival." ALA Notable Book.

• • •

SCHOOL LIBRARY JOURNAL says of her book ***Talk: Teen Art of Communication***, "Explores philosophical and psychological aspects of communication, encourages young people." The ***Teen Brain Book: Who and What Are You***, "Practical and scientifically-based guide to the teen brain." ***In and Out of Your Mind: Teen Science***, "Thought-provoking guide into the mysteries of inner and outer space." New York Public Library Best Books for Teens.

• • •

THE NEW YORK TIMES BOOK REVIEW says of Carlson's YA Science Fiction novel ***The Mountain of Truth***, "Anyone who writes a teenage novel that deals with a search for the Truth must have a great respect for the young…speaks to the secret restlessness in the adolescent thinker." ALA Notable Book.

• • •

KIRKUS REVIEWS YA novel ***The Human Apes***, "Vastness of this solution to the human dilemma." ALA Notable Book.

PUBLISHERS WEEKLY says of ***Where's Your Head? Psychology for Teenagers***, co-authored by Dale Carlson and Hannah Carlson, M. Ed., C.R.C. "Psychological survival skills…covers theories of human behavior, emotional development, mental illnesses and treatment." Christopher Award.

• • •

THE MIDWEST BOOK REVIEW says of ***Who Said What: Philosophy for Teens***, "Evocative, thought-provoking compilation and very highly recommended reading for teens and young adults." VOYA Honor Book.

• • •

SCHOOL LIBRARY JOURNAL says of ***Stop the Pain: Teen Meditations***, "Much good advice is contained in these pages." New York Public Library Best Books for Teens.

• • •

SCHOOL LIBRARY JOURNAL'S Dodie Ownes says of ***Are You Human, or What?*** "Carlson makes a unique addition to help teens understand themselves…A primer of evolutionary psychology, this will not only be useful to teens, it is recommended for high school libraries and counseling offices, young adult collections in public libraries, and community college libraries."

# COSMIC CALENDAR

## The BIG BANG to Your Consciousness

# COSMIC CALENDAR
## The BIG BANG to Your Consciousness

DALE CARLSON

Edited by Kishore Khairnar, M.S. Physics
Illustrations By Nathalie Lewis

**Nathalie Lewis
Graphic Book**

BICK PUBLISHING HOUSE     MADISON CT     2009

Text © 2009 by Dale Carlson
Illustrations © 2009 by Nathalie Lewis

Edited by Director Editorial Ann Maurer
Senior Science Editor Kishore Khairnar M.Sc. Physics

Book Design by Jennifer A. Payne, Words by Jen
Cover Design by Greg Sammons

**www.bickpubhouse.com**

Library of Congress Cataloging-in-Publication Data

Carlson, Dale Bick
    Cosmic calendar: from the big bang to your consciousness/Dale Carlson and Kishore Khairnar; illustrations by Nathalie Lewis.
    p.cm.-(Graphic nonfiction series)
    Includes bibliographical references and index.
    ISBN 978-1-884158-34-6 (quality pbk: alk.paper1. Science-Popular works. I Khairnar, Kishore, II. Title.

Q162.C356 2009
500-dc22                                                        2008047656

AVAILABLE THROUGH:
- Distributor: BookMasters, Inc., AtlasBooks Distribution,
  Tel. (800) BOOKLOG, Fax: (419) 281-6883
- Baker & Taylor Books
- Ingram Book Company
- Follett Library Resources, Tel: (800)435-6170 Fax: (800) 852-5458
- Amazon.com

Or: Bick Publishing House
307 Neck Road
Madison, CT 064433
Tel: (203)245-0073 Fax: (203) 245-5990

Printed by NcNaughton & Gunn, Inc. USA

IN DEDICATION

To Jennifer A. Payne, Words by Jen

With love and gratitude for superb book and graphic design.

## ACKNOWLEDGMENTS

To Kishore Khairnar, M.S. Physics, friend, philosopher, and gold medal physicist, without whose wisdom and editorial insights I could not have written this book.

To Will Corbin 17, Eric Fox 16, Alyssa Fox 16, Andrea Rivera 17—four teen editors without whose constructive help this book could have been in trouble.

To Hannah Carlson, M.Ed., LPC., Senior Director, Dungarvin, who, after decades of the same best schools as myself, graduated as undereducated in physics and chemistry as I did, and who learned, with me, enough to read and reread this book in manuscript.

To Ann Maurer for her years of patient editorial direction, devotion, and support.

To all the scientists and science writers who have educated us and made us at least scientifically literate enough to participate in necessary decisions for ourselves and the planet.

# BOOKS BY DALE CARLSON

TEEN FICTION:
*Baby Needs Shoes*
*Call Me Amanda*
*Charlie the Hero*
*The Human Apes*
*The Mountain of Truth*
*Triple Boy*

• • •

TEEN NONFICTION:
*Are You Human or What? Teen Psychological Evolution*
*Girls Are Equal Too: The Teenage Girl's How-to-Survive Book*
*In and Out of Your Mind: Teen Science, Human Bites*
*Stop the Pain: Teen Meditations*
*TALK: Teen Art of Communication*
*The Teen Brain Book: Who and What Are You?*
*Where's Your Head?: Psychology for Teenagers*
*Who Said What? Philosophy Quotes for Teens*

• • •

ADULT NONFICTION:
*Confessions of a Brain-Impaired Writer*
*Stop the Pain: Adult Meditations*

• • •

WITH HANNAH CARLSON

*Living with Disabilities: 6-Volume*
*Basic Manuals for Friends of the Disabled*

*The Courage to Lead: Start Your Own Support Group*
*Mental Illnesses & Addictions*

• • •

WITH IRENE RUTH
*First Aid for Wildlife*
*Wildlife Care for Birds and Mammals:*
*7-Volume Basic Manuals Wildlife Rehabilitation*

# TABLE OF CONTENTS

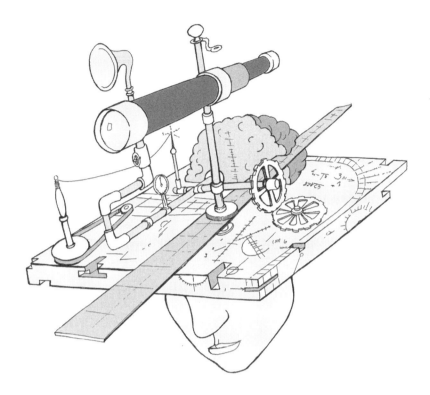

# FOREWORD

From the original edition, *In and Out of Your Mind: Teen Science, Human Bites*

"I want to know how God created this world. I am not interested in this or that phenomenon, the spectrum of this or that element. I want to know His thoughts. The rest are details," said Albert Einstein, who laid the groundwork for the twentieth century's two fundamental theories: general relativity and quantum theory.

Yet the great physicist, like his successor, theoretical physicist Stephen Hawking, sought not only science's holy grail, the Theory of Everything, but what Hawking calls 'the mind of God.'

Scientists, like the rest of us, want to know how this world got created, and if there is something called God behind this creation. The interesting question is, how do we find out? Through the scientific spirit? Through the religious spirit? Or are these, in fact, the same?

Science as it is taught today is usually based on a fragmentary approach: physics, chemistry, biology, mathematics, and so forth. This division of information is necessary to study details. But without grasping the whole vision of science, the study of details becomes a rather superficial activity. Science then becomes something to be studied in the laboratory only and appears to be far removed from our day-to-day life. Does science hold something greater, which as human beings should interest each one of us? Does it hold the answers to those deeper questions humans have forever asked? How did the universe come into being? Who are we? Where did we come from? Is there God?

It is only recently that we have used science primarily instead of religion to explore these questions. According to science, the story of creation started with the big bang, about fifteen million years ago. This super explosion led to the formation of galaxies, stars, the planetary systems around stars. On the planet Earth, favorable conditions led to the creation of living forms. These evolved into us, human beings. In this book, the author Dale Carlson has unfolded the story of how we have come to be, what we are in terms of various branches of science.

- The big bang that led to the formation of various elements and the physical laws that govern the universe are the specialized study of the science called physics.

- The elements thus created combined to form molecules of various substances: the study of these substances and their properties is called chemistry.

- Various chemical molecules combined to create most complex organic structures. When we study the evolution of these into living forms, we call the science biology.

- Among these living forms, the human has evolved with the ability to think. This ability to think logically, to rationalize, to observe, to learn, to pass on knowledge and culture, has brought into existence an altogether new world: the inner, psychological world of human beings. The study of this world and its phenomena has created yet another branch of science—psychology.

Among the objectives of psychology are to bring us more knowledge and understanding about ourselves, about how to manage our lives so that we suffer and make others suffer less. Insecurity and habit have encouraged us to listen to external teachers, psychotherapists, priests, gurus for information about ourselves. J. Krishnamurti, the great philosopher and teacher points out that each of us can learn about ourselves simply by observing ourselves in relationship with other human beings, in our daily thoughts, feelings, reactions, behavior. Looking at ourselves in the mirror of relationship is an ongoing inquiry that combines compassion and freedom from authority in the purest of both religious and scientific spirit—not according to some established religion or to some established scientific theory. Beliefs have no place in scientific inquiry, or in relieving psychological suffering. Only with sensitive awareness and observation can we learn anything new.

In science education, developing the scientific spirit is far more important than to cram the brain with the details of information. The scientific spirit is the spirit of precision, accuracy, and efficiency, not personal or national idiosyncrasy or bent according to any organized religion. It is also the spirit of adventure. You follow facts, not individuals. You don't worship the lamppost, you use the light to go on. To develop such a spirit should be the aim of science education. This means emphasis needs to be given in the arts of observation, experimentation, and learning for oneself. And once developed, these arts, this same independent spirit can be applied to learning about the inward life as well as the outer world. In this way, the spirit of science and the spirit of religious compassion become one. This is far more important than the mere discovery of yet more scientific facts, lest these be used only for self-centered purposes.

Thus the scientific spirit, when applied to learning about ourselves, our relationships, creates as well the true religious spirit. In the words of J. Krishnamurti, "A religious man is one who is helping to free the individual, and himself, from all the cruelty and suffering in life—which means that he is free from all belief. He has no authority, he does not follow anyone, because he is a light unto himself, and that light arises from self-knowledge. It is the liberation that comes into being when the individual completely understands himself. The religious man is one who is creative, not in the sense of painting pictures or writing poetry, but there is in him a creativity, which is everlasting, timeless."

I think the aim of science education is to bring about this kind of religious mind.

— *Kishore Khairnar, M.S., Physics*

# INTRODUCTION

Intelligent people ask, and have been asking for thousands of years: where did we come from? where are we going? why are we here? are we alone in this great, vast universe? and is anybody up there in charge?

What is interesting is why the human race asks such questions, why we don't just live instead of trying to find answers to the mysteries of life, death, and the cosmos. This curious, questing intelligence of ours is as great a mystery as life itself. It's where intelligence and confusion fuse; from Confucius to Socrates, you, and me, this question of our consciousness itself still remains unanswered.

There seems to be a wide gap in explanations about human consciousness, about our inner lives, our mental lives. There are arguments that lean toward nature and those that lean toward nurture, between the physiological and psychological, between the biological and cultural/environmental explanations of the evolution of our brains and our behaviors.

Scientists like Stephen Jay Gould describe the argument as the difference between biological determinism and biological potential. Philosophers like J. Krishnamurti discuss the difference between thought, or intellect, which is the past, the known;

and intelligence, the human capacity for new perception. What is clear is that we are a species that not only behaves, but is aware of its own behavior, that doesn't just live but wants to understand the wherefrom's, the why's, and the whatfor's of its lives. And we want to know about the seeming difference between us and other animals, if, that is, there is a difference. What seems unique to humans is what we call conscious awareness. We want to know why we are conscious, and we also want to know whether that consciousness is on its own, or linked to a universal consciousness.

We want to know about the natural history of our world, as well as human history, about the origin of life and, even more deeply, what constitutes life itself. Physics and molecular biology indicate that all mass is energy, all energy is mass, and at the micro level, a quark is a quark in your atoms, my atoms, and the atoms of a star. That being so, isn't a stone alive, since there is the same movement in its atoms as in ours? For the same reason, isn't a chicken as important as I am? Gould says life is a bush, with the human race a mere twig—not a tree with us at its glorious summit. I'm with Gould on that, if for no other than aesthetic reasons. Look in the mirror, and you will see a naked, hairless creature with brief bushes to cover vital parts, flightless by bird standards, puny in proportion to the strength of any decent-sized mammal, unable to swim at all well by most undersea measurements, slow on foot, helpless against the weather, uncamouflaged, unable to burrow, smell, or taste airborne scents, or even entirely wean its young for a couple of decades. We don't even have numbers on our side: bacteria constitute eighty percent of life on earth. The one survival tool we have is two or three pounds of nerves and tissue between our ears—and if anything damages that, we're dead in the water.

But that is what it seems to come down to—our perceived difference from all we survey and measure ourselves against: the human brain.

So, some more questions arise.

If our brains are so special, why do they drive us crazy, make us suffer so much anguish and anxiety, feel so alone despite our numbers and proximity (there are six billion of us, often pushed in together)? Why don't we see that our thoughts interfere with our perception of reality? Why do we try to reproduce ourselves robotically, search for alien life elsewhere in the universe, call out to a night full of stars the unknown name of an unknown God!

If our brains are so special, whey can't they solve their own problems? Why can't our brains even look at themselves and their own processes with any degree of psychological clarity?

Why are we such a danger to ourselves? A chemistry set, a reasonable knowledge of biology, or even just enough pilot training to hijack a plane and crash it into a chosen site as terrorists did to New York City and Washington in September 2001, and someone can plunge the world into war. High tech nuclear missiles aren't necessary: only suicidal hate and an inhuman lack of ethics.

Is it possible we're so smart we've got everything all wrong? Have these brains of ours gotten so complex they're missed the simple truths, the obvious answers?

The trouble is, we see everything we can see with these senses, these brains. If these instruments are skewed in any way, we may be missing the whole, beautiful, unified truth. Physicist Paul Davies, in his book *The Fifth Miracle*, thinks there are still undiscovered laws of physics at work in a biofriendly universe, that the origin of life is discoverable, and furthermore, that we are not alone out there in the dark.

— *Dale Carlson*

# CHAPTER 1
# Are We the Aliens?

# WHERE, AND WHAT, IS THE ORIGIN OF LIFE?

OKAY, OKAY, SO STARS AND DUST EXPLODED FROM THIS COSMIC BALL—WHAT ABOUT ME! WHAT ABOUT MY CAT?

I think they're asking how come they're alive and not just another rock?

Yeah, where did life come from?

CAT? WHAT ABOUT ME, YOUR FRIEND, HERE?

I wanna get the genius who thought up me and my life getting stuck in my teens for a decade!

NO ONE KNOWS YET FOR SURE EXACTLY WHERE LIFE CAME FROM OR HOW IT BEGAN. SCIENTISTS HAVE 3 THEORIES.

LIFE MAYBE BEGAN IN A WARM POND ON EARTH'S SURFACE, AS DARWIN SUGGESTS.

OR MAYBE IT ERUPTED FROM UNDER THE EARTH'S CRUST.

OR IT ARRIVED IN SOME METEORITE THAT CRASHED INTO EARTH'S SURFACE FROM OUTER SPACE.

PHYSICIST AND PHILOSOPHER IMMANUEL KANT OPENED THE WAY FOR EINSTEIN BY POINTING OUT OUR A PRIORI, INBORN WAY OF SEEING THINGS IN 4-SPACE.

Height
Depth
Width
Time

**3.** Humans also have another point of view problem, a 5th dimension, partly inborn, partly cultural, called: ME!!!! !!!!! SELF!!!!

ME

Thought itself actually invents a little 'me' to give itself the continuity and security it needs.

J. KRISHNAMURTI, PHILOSOPHER & TEACHER

There is no little 'me', no 'I', no homunculus inside our brains. We invent a 'self' because we're afraid of being nothing, of having no thinker in our heads. Our thoughts invent a 'self' out of our memories because we think having a 'self' protects our bodies.

The trouble is, our 'self', our own memories and cultural/personal conditioning colors and distorts everything, including our own thinking.

ALEX COMFORT, NEUROPSYCHIATRIST

7

BUT GAGGLES OF PARTICLE PHYSICISTS, PSYCHOLOGISTS, AND PHILOSOPHERS HAVE SHOWN OTHER WAYS THAT THINKING CAN TAKE PLACE WITHOUT A SELF OR THINKER.

THERE IS AN INTELLIGENCE THAT DOES NOT PROCESS INFORMATION ACCORDING TO 'I THINK'. IN HUMAN BRAINS, THINKING CAN TAKE PLACE WITHOUT THE SELF GETTING IN THE WAY AND COLORING THE TRUTH WITH ITS OWN OPINIONS AND PROBLEMS. THIS IS JUST TRUE OBSERVATION ITSELF. AND NOT ONLY IN SCIENCE, BUT IN LIFE.

Humans have the capacity to use intelligence, direct observation to see the truth of something, including one's own thoughts and feelings.

Intellect and knowledge have their place, to use language, to remember facts, to function practically in the world. But to see something new, scientifically or psychologically, intellect's memory just gets in the way of new observations.

J. KRISHNAMURTI, PHILOSOPHER & TEACHER

In science, we use demonics. A demon is a non-positional 'I', a logic that does not depend on our limited 5 dimensions.

ALEX COMFORT, NEUROPSYCHIATRIST

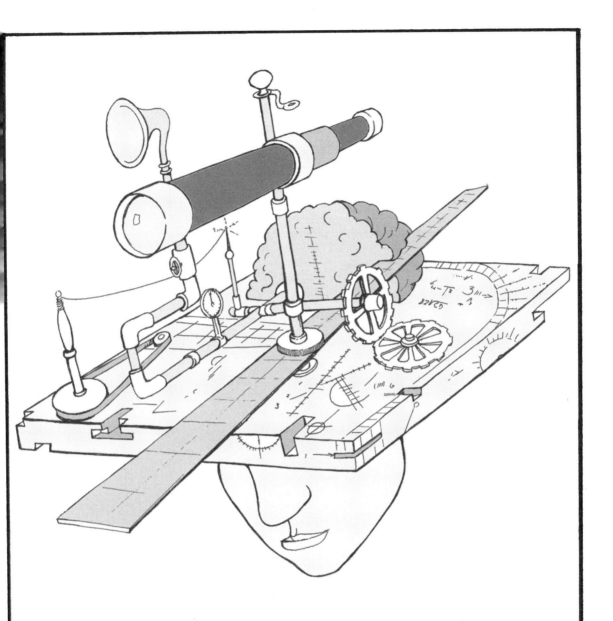

| OBSERVATION | MEMORY |
| INSIGHT | KNOWLEDGE |
| NON-POSITIONAL 'I' ⟶ **VS** ⟵ | 4-DIMENSIONS-ONLY |
| INTELLIGENCE: INSIGHT, FRESH, NEW OBSERVATION OF WHAT ACTUALLY IS | INTELLECT: MEMORY, KNOWLEDGE, THE PAST |

LISTEN UP! THIS IS WHAT MY MOTHER WAS GOING ON ABOUT WHEN I ASKED WHERE MY BABY DOLL CAME FROM.

THEY DON'T LISTEN TO QUESTIONS RIGHT.

# WHERE DID LIFE COME FROM?

15 BILLION YEARS AGO, the ball of everything, all matter, exploded into hot energy.

FIRST SPLIT SECOND, energy converted itself into all the basic physical laws, particles, and in the universe—protons, neutrons, and electrons.

THREE MINUTES LATER, these formed the nuclei of atoms.

ONE BILLION YEARS LATER, clumps of hydrogen and helium atoms formed, each destined to become some huge cosmic body like a galaxy.

ABOUT FOUR BILLION (3.8) YEARS AGO, billions of galaxies full of stars and solar systems like ours formed—and it was cool enough for LIFE TO FORM, for INERT chemicals to get ERT.

YEAH? HOW?

THEY HAVEN'T TOLD ME YET.

WHAT I STILL WANT TO KNOW IS, IF LIFE CAME TO EARTH ON A CRASHING ASTEROID, ARE WE THE ALIENS?

# WHAT IS LIFE?

WADDYA MEAN THE LINE BETWEEN ME AND THAT CAT IS FUZZY?

WADDYA MEAN THE LINE BETWEEN ME AND THAT GIRL IS FUZZY?

Views among battalions of chemists, astrophysicists, evolutionary biologists vary about when chemistry turns into life. After all, everything is made of atoms, and atoms are exchanged continually throughout the biosphere, in and out of living organisms, all the time.

You, the table, a mouse, and the Sun are all made out of the same stuff—atoms.

AND EVERYTHING IN THE UNIVERSE IS, AFTER ALL—STARDUST!

THEY'RE TELLING US YOU AND ME AND EVERYTHING ELSE IS MADE OF THE SAME STUFF.

AT LEAST TELL ME I'M NO SOCCER BALL.

I can give you that one. If I throw you both against the wall, the effect is different. Also, even if you're both made out of atoms, you and the soccer ball...

...your soccer ball does not have a life of its own. You do. For a soccer ball to move, you have to kick it. You move all by yourself.

PAUL DAVIES

13

# THE QUALITIES THAT DETERMINE LIFE ARE:

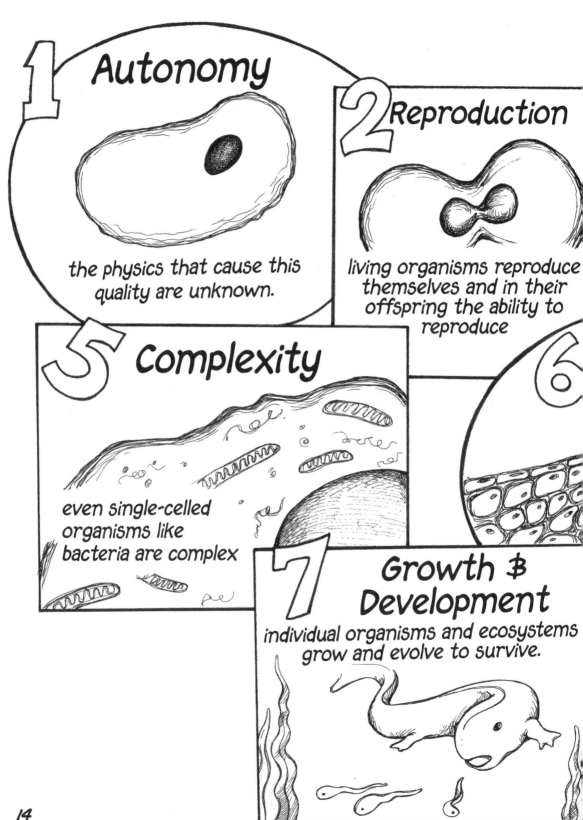

**1** Autonomy

the physics that cause this quality are unknown.

**2** Reproduction

living organisms reproduce themselves and in their offspring the ability to reproduce

**5** Complexity

even single-celled organisms like bacteria are complex

**6**

**7** Growth & Development

individual organisms and ecosystems grow and evolve to survive.

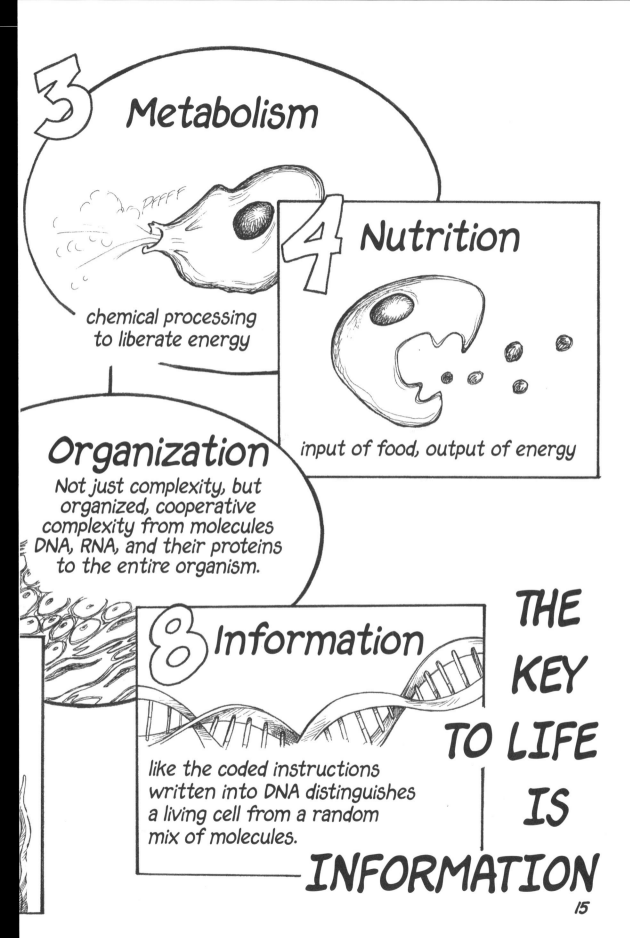

**3 Metabolism**

PFFFF

chemical processing
to liberate energy

**4 Nutrition**

input of food, output of energy

**Organization**

Not just complexity, but
organized, cooperative
complexity from molecules
DNA, RNA, and their proteins
to the entire organism.

**8 Information**

like the coded instructions
written into DNA distinguishes
a living cell from a random
mix of molecules.

THE
KEY
TO LIFE
IS
INFORMATION

# LIVING CELLS

A LIVING CELL IS THE CHEMISTRY OF ITS CONTENTS: THE NUCLEIC ACIDS DNA AND RNA, AND THEIR HARDWORKING AGENTS, THE PROTEINS. DNA HAS THE INSTRUCTIONS, RNA TRANSCRIBES AND RELAYS DNA INSTRUCTIONS TO THE PROTEINS.

IT'S LIKE A COMPUTER. PROTEINS ARE THE HARDWARE. THE BASE PAIRS OF THE DNA ARE THE CELL'S SOFTWARE.

Life is information technology written small.

Our ability to last is based on our ability to change. We must adapt or die. Without variation, adaptation is impossible, and the genes will die out.

PAUL DAVIES

To survive is to change in response to what is going on in your environment.

DR. FRANKENSTEIN SAID LIFE CAME FROM LIGHTNING.

SO THERE'S NO SIMPLE ANSWER TO WHAT IS LIFE. LIFE INVOLVES PRIMARILY REPRODUCTION AND METABOLISM. AND WHAT ELSE SETS IT APART FROM NONLIVING MATTER IS THE ABILITY TO TAKE IN INFORMATION TO EVOLVE, TO CHANGE AND ADAPT TO CHANGING ENVIRONMENTS.

I THINK LIFE COMES FROM THE CAT GOD.

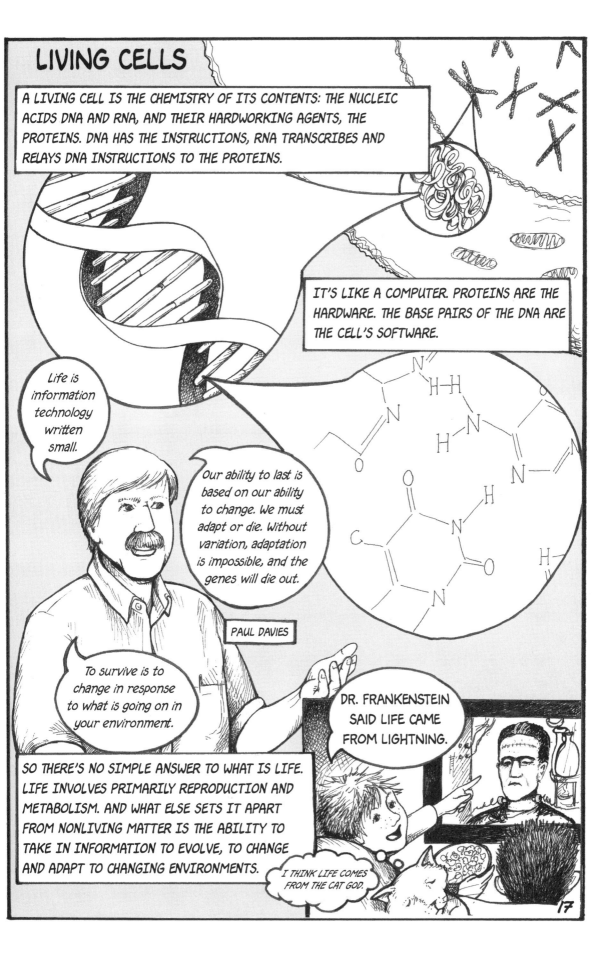

# CHAPTER 2
## Your Evolution Into Present Standard Operational Equipment

# A TEENAGERS COSMIC CALENDAR

EVOLUTIONARY BIOLOGIST CARL SAGAN WROTE DOWN A FAMOUS COSMIC CALENDAR IN HIS BOOK *DRAGONS OF EDEN—A CHRONOLOGICAL HISTORY OF THE UNIVERSE FROM THE BIG BANG ABOUT 15 BILLION YEARS AGO*, TO EARTH'S BIRTH, TO HUMAN HISTORY.

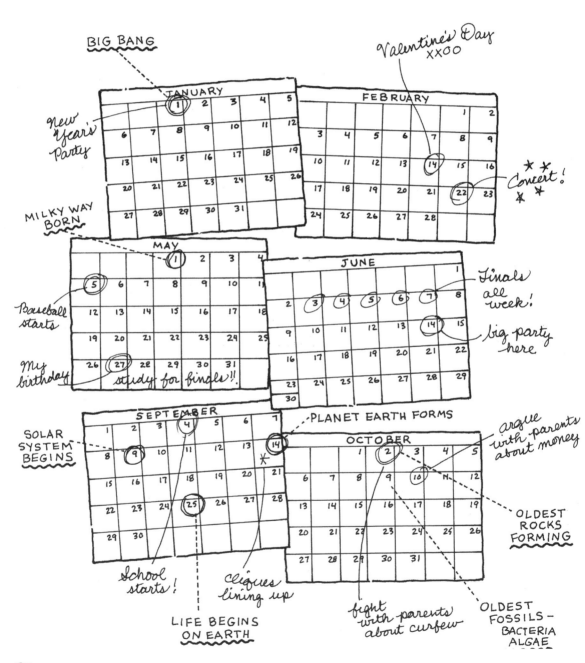

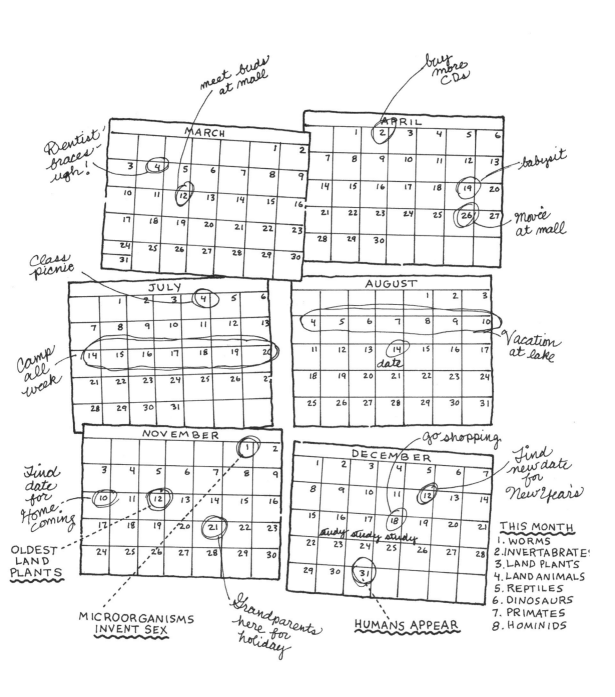

IF 15 BILLION YEARS OF COSMIC HISTORY IS REPRESENTED BY ONE YEAR, OUR WHOLE HUMAN HISTORY OCCURS IN THE LAST 10 HOURS!

* Cosmic Calendar illustration by Carol Nicklaus.

# FREAKS & NERDS

MY DEFINITION OF A FREAK IS A BRAIN THAT CAN'T HELP SEEING THE TRUTH AND SPEAKING UP ABOUT IT, RIGHT THROUGH ITS OWN INHERITED KNOWLEDGE AND CULTURAL MUD.

Contrary to the beliefs of jocks, sports fans, soccer moms and ice hockey dads, human bodies ain't so much.

PULITZER PRIZE WINNER NEUROSCIENTIST, PSYCHOLOGIST, SCIENCE WRITER EXTRAORDINAIRE STEVEN PINKER SAYS IN HIS BOOK *HOW THE MIND WORKS*...

"Human evolution is the original revenge of the nerds." We have no poisonous fangs, no night vision to speak of, no great physical strength or speed, comparatively slow physical reflexes, no useful tails, sensitive noses or trunks, or flight feathers.

True, we have upright posture, frontal vision and thumbs for precision manipulation.

But actually, it is not our bodies but our BRAINS "our behavior and the mental programs that organize it" that allows us "humans control of the fate of tigers, rather than vice versa."

WHAT PINKER IS SAYING IS THAT THE NERDS OF THIS EARTH, NOT THE JOCKS, ARE IN CHARGE. SO WE BETTER UNDERSTAND OUR HUMAN INTELLECTS AS WELL AS OUR EVOLUTIONARY BIOLOGICAL AGENDA. INTELLECT, IN OUR SPECIES, IS OUR GREATEST ASSET. IT IS ALSO OUR MOST DANGEROUS.

# WHAT IS EVOLUTION?

CHARLES DARWIN SAYS IN HIS ORIGIN OF SPECIES: THIS PROCESS IS WHAT I CALL NATURAL SELECTION.

**1** All organisms tend to produce more offspring than can possibly survive.

**2** Offspring vary—they are not carbon copies.

**3** Some variations are inherited

**4** Many offspring die: the survivors are those who adapt best to their changing local environment—and their characteristics will be passed on to future generations.

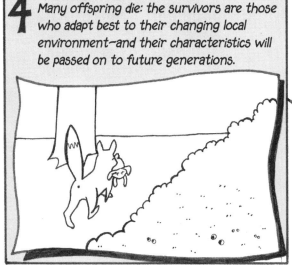

Adaptive heredity does not mean progress. "Humans are just another twig on the bush of life," not the top of a tree. If numbers count, then this is the Age of Bacteria, not us.

STEPHEN JAY GOULD

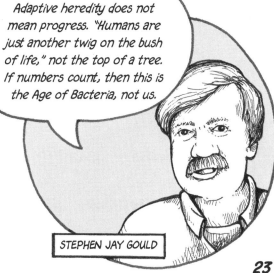

23

ACCIDENTS HAVE ALSO AFFECTED WHO IS OR ISN'T THE EARTH'S BIG SHOT. PALEONTOLOGISTS, WHO STUDY PAST LIFE FROM FOSSIL REMAINS, POINT OUT MASS EXTINCTIONS OF SPECIES.

If an asteroid hitting Earth had not cooled the climate and wrecked the food supplies for the big dinosaurs (we still have small, flying dinosaurs, only we call them birds)...

small mammals like us could never have made it.

SIGMUND FREUD REMINDS US TO STOP THINKING WE HUMANS ARE THE CENTER, THE TOP, THE POINT OF ALL CREATION.

All major evolutions in science...end in the dethronement of human arrogance. We use to think Earth was the center of the universe until Copernicus, Galileo, and Newton and modern astrophysicists showed us Earth as a tiny satellite of one of millions of stars in one of millions of galaxies in one of who knows how many universes.

OOPS!

STAND IN A CORNER AND COUNT YOURSELF. YOU'RE NOT SO MANY.

# HUMAN EVOLUTION: FIRST, EARLY & LATE-BLOOMING PEOPLE

THERE IS ONLY ONE CURRENT HUMAN RACE (SPECIES). THERE ARE ONLY GENETIC VARIATIONS IN HUMANS, NOT DIFFERENT RACES.

THERE ARE LOTS OF DOG SPECIES.

TOLD YOU CATS WERE LIKE HUMANS—LOTS OF COLORS AND SIZES, BUT ONLY ONE SPECIES.

Skin tones depend on how much melanin is needed for protection from local doses of sunlight—melanin makes the skin darker. A lot is needed in sunny places, not much in Lapland.

Slight differences in a) bone structure—African thigh bones are longer for running through jungle terrain, so legs can carry them where cars can't go...

Isolation or intermarriage can make certain features more dominant.

b) eyelid structures and nose shapes vary depending on vicious desert winds or really moist climates or colder seasonal air: the Mongolian eye-fold protects the eye from winds; narrow noses protect from cold air, wider noses let in more oxygen in hot, moist climates.

25,000 YEARS AGO HOMO SAPIENS BEAT OUT THE NEANDERTHALS. HOMO SAPIENS IS THE ONLY HOMINID STILL WALKING THE EARTH.

NO WONDER WE FEEL LONELY AND SUPERIOR...

# EARLY US

So here we are.

CARL SAGAN SAYS OUR ANCESTORS STOOD UP AND WALKED ON 2 LEGS BEFORE THEY EVOLVED BIG BRAINS. PHYSICAL CHANGES LIKE UPRIGHT POSTURE CHANGES THE NEURONAL PATHWAYS OF THE BRAIN.

I think it was the African teenager Lucy (we have discovered her bones) 3 1/2 million years ago who changed human brains forever, when she carried her baby out of the jungle onto the grassy savannah to check out the neighborhood for lions and/or food.

Humans like Lucy left the shrinking jungle to look for food out on the grassy savannah. Standing upright helped her look out for food and predators, and freed her hands. This changed her brain and all our brains from then on. Frontal vision helped her remember images of dangerous things and good food. Memory and image-making led to language and culture and technology and passing it all on in art and writing. Frontal vision also changed sex into relationships.

TOLD YOU IT WAS A TEENAGE GIRL WHO INVENTED HUMAN BEINGS.

WE HAD TO LOOK EACH OTHER IN THE EYE.

A GIRL, AND HER CAT.

# DATES FOR THE APPEARANCE OF HOMINIDS

*HERE ARE SOME APPROXIMATE DATES FOR THE APPEARANCE OF HOMINIDS.*

4.2 *mya*

*Ardipithecus ramidus:* in Ethiopia, the earliest hominid biped, apelike but upright.

4.4 *mya*

*Australopithecus anamensis:* in Kenya, definitely crossed the line from ape to human.

2.5 to 1.8 *mya*

The robusts, *Paranthropus boisei, P. robustus* and others made Africa host to multiple kinds of early people.

2 to 1.4 *mya*

Fossils from this period announce the earliest appearance of the genus *Homo,* among them *Homo habilis* (our early tool man or handywoman) and *Homo ergaster,* the first modern body form and probably the first in the exodus to other continents, Europe and reaching China and Java by about 1.8 mya.

200,000 to 30,000 years ago

Neanderthal people may have coexisted with us from Europe to Asia. They used tools and language, buried their dead, supported their old and sick. They may not have been directly our ancestors, but a divergent line from 500,000 years earlier.

*MYA = MILLION YEARS AGO

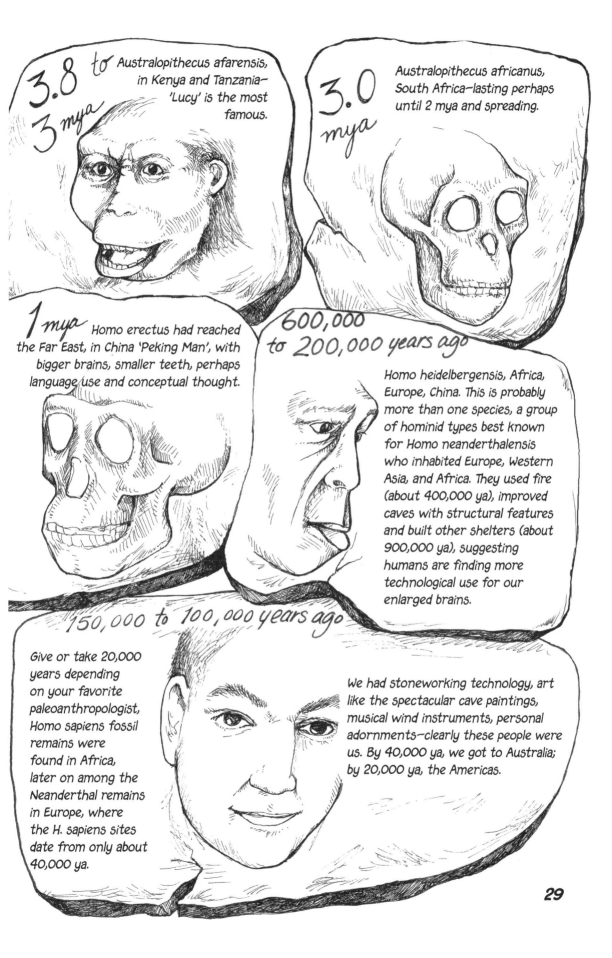

**3.8 to 3 mya** Australopithecus afarensis, in Kenya and Tanzania—'Lucy' is the most famous.

**3.0 mya** Australopithecus africanus, South Africa—lasting perhaps until 2 mya and spreading.

**1 mya** Homo erectus had reached the Far East, in China 'Peking Man', with bigger brains, smaller teeth, perhaps language use and conceptual thought.

**600,000 to 200,000 years ago** Homo heidelbergensis, Africa, Europe, China. This is probably more than one species, a group of hominid types best known for Homo neanderthalensis who inhabited Europe, Western Asia, and Africa. They used fire (about 400,000 ya), improved caves with structural features and built other shelters (about 900,000 ya), suggesting humans are finding more technological use for our enlarged brains.

**150,000 to 100,000 years ago** Give or take 20,000 years depending on your favorite paleoanthropologist, Homo sapiens fossil remains were found in Africa, later on among the Neanderthal remains in Europe, where the H. sapiens sites date from only about 40,000 ya.

We had stoneworking technology, art like the spectacular cave paintings, musical wind instruments, personal adornments—clearly these people were us. By 40,000 ya, we got to Australia; by 20,000 ya, the Americas.

WE HAVE NO DIRECT CLUES ABOUT THE INTERACTIONS BETWEEN THE NEANDERTHALS AND HOMO SAPIENS, BUT FROM THE RECORD OF OUR BEHAVIOR SINCE, WE CAN IMAGINE WHAT WE DID TO THE NEANDERTHAL COUSINS. EVEN THOUGH THERE ARE SOME THEORIES ABOUT INTERBREEDING, AND THOUGH DISCOVERIES OF VARIOUS HOMINIDS STILL OCCUR THROUGHOUT THE WORLD, WHAT WE DO KNOW IS THAT OUR SPECIES BECAME UTTERLY INTOLERANT OF COMPETITION. WE SIMPLY DO NOT ALLOW IT.

A NOTE: WE STILL DO NOT KNOW WHEN OR HOW OUR LINEAGE DEVELOPED SYMBOLIC THOUGHT, HOW THE MODERN HUMAN BRAIN CONVERTS CHEMISTRY AND NERVE CONNECTIONS IN THE BRAIN INTO CONSCIOUSNESS. IT MAY HAVE BEEN A CULTURAL STIMULUS, SUPPORTED BY LANGUAGE AND IMAGE-MAKING, BUT WHATEVER ITS ORIGIN, CONSCIOUSNESS HAS MADE OF US, THE NERDS OF THE EARTH AND THE BRAINIEST OF ALL SPECIES, THE MOST POWERFUL ANIMAL, THOUGH NOT NECESSARILY THE MOST INTELLIGENT CREATURE, ON OUR PLANET.

# CHAPTER 3
# Your Physical Universe: All the Matter There Is

## THE BIG, VISIBLE, MACROWORLD OF THE UNIVERSE

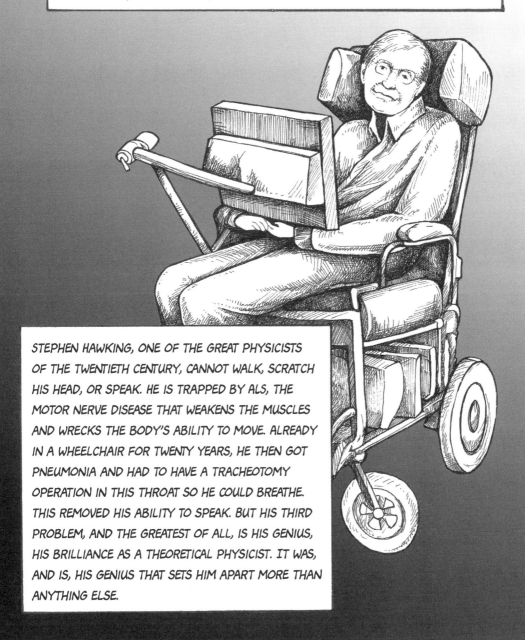

STEPHEN HAWKING, ONE OF THE GREAT PHYSICISTS OF THE TWENTIETH CENTURY, CANNOT WALK, SCRATCH HIS HEAD, OR SPEAK. HE IS TRAPPED BY ALS, THE MOTOR NERVE DISEASE THAT WEAKENS THE MUSCLES AND WRECKS THE BODY'S ABILITY TO MOVE. ALREADY IN A WHEELCHAIR FOR TWENTY YEARS, HE THEN GOT PNEUMONIA AND HAD TO HAVE A TRACHEOTOMY OPERATION IN THIS THROAT SO HE COULD BREATHE. THIS REMOVED HIS ABILITY TO SPEAK. BUT HIS THIRD PROBLEM, AND THE GREATEST OF ALL, IS HIS GENIUS, HIS BRILLIANCE AS A THEORETICAL PHYSICIST. IT WAS, AND IS, HIS GENIUS THAT SETS HIM APART MORE THAN ANYTHING ELSE.

WHEN HE WANTED TO GET MARRIED, STEPHEN HAWKING, LIKE ANY OTHER MAN, HAD TO GET A JOB. TO GET A JOB, HE HAD TO GET A DEGREE, TO GET A DEGREE IN HIS FIELD HE HAD TO WRITE A THESIS, AND IN THE COURSE OF HIS THESIS, HE DID THE IMPOSSIBLE. HE BROUGHT TOGETHER THE TWO GREAT—AND CONTRADICTORY—THEORIES OF THE CENTURY: the theory of relativity

and quantum theory, the strange and unexpected realities of the small, the subatomic world of particles invisible to the naked eye

HIS POINT WAS THAT WHILE EINSTEIN'S MATHEMATICS PREDICTED THE EXPANSION OF THE UNIVERSE, AS WELL AS THE FORCE OF GRAVITY THAT PULLED MATTER TOGETHER INTO STARS AND GALAXIES, THERE WAS STILL A PROBLEM IN UNDERSTANDING THE PRIMEVAL ATOM. THIS WAS ONLY SOLVABLE THROUGH QUANTUM PHYSICS.

Hawking described the perfect model for the explosion of all the collapsed matter in the universe we call the big bang.

THEN HAWKING HAD A FOURTH PROBLEM. HE FELT IT WAS IMPORTANT FOR ALL OF US, NOT JUST A FEW SCIENTISTS, TO UNDERSTAND THE UNIVERSE WE LIVE IN, BE EQUALLY RESPONSIBLE WITH SCIENTISTS AND POLITICIANS FOR DECISIONS MADE THAT AFFECT EVERY LIFE IN THE UNIVERSE, AS WELL AS TO TAKE PART IN THE DISCUSSION OF WHY WE AND THE UNIVERSE EXIST. HE WANTED TO EXPLAIN IN UNDERSTANDABLE TERMS, WITHOUT THE MATHEMATICS WHICH IS THE NORMAL LANGUAGE OF PHYSICISTS, THE ENTIRE UNIVERSE—ITS ORIGIN, ITS FATE, HOW IT WORKS BOTH AT THE LEVEL OF LARGE VISIBLE OBJECTS LIKE STARS AND GALAXIES OF STARS, AND AT THE INVISIBLE LEVEL OF ATOMS AND THE SUBATOMIC WORLD OF QUARKS AND QUANTA.

Wait a minute—this guy uncovered the secrets of the universe all because he wanted to get married!?

I told you hooking up with me could be the greatest thing that ever happened to you.

THE USE OF A MOTORIZED WHEELCHAIR ON WHICH IS MOUNTED A SMALL PERSONAL COMPUTER AND A SPEECH SYNTHESIZER GIVES STEPHEN HAWKING THE NECESSARY MOBILITY AND A SYSTEM OF COMMUNICATION. THAT SOLVED THE FIRST TWO PROBLEMS. HE WROTE HIS THESIS AND GOT A JOB, NOW HOLDING NEWTON'S CHAIR AT CAMBRIDGE UNIVERSITY, AND, SOLVING HIS THIRD PROBLEM, IS NOW HAPPILY MARRIED.

TO SOLVE THE FOURTH PROBLEM, HE WROTE A POPULAR BOOK, *A BRIEF HISTORY OF TIME FROM THE BIG BANG TO BLACK HOLES*, WITHOUT MATH EQUATIONS, SO EVEN PEOPLE LIKE ME WITH NO SCIENTIFIC EDUCATION COULD UNDERSTAND THE FIELDS OF COSMOLOGY (ASTROPHYSICS, THE STUDY OF THE GREATER UNIVERSE) AND QUANTUM THEORY (PARTICLE PHYSICS, THE STUDY OF THE VERY SMALL OBJECTS). HE EVEN EXPLAINS HIS GROUND-BREAKING RESEARCH INTO BLACK HOLES, CLUES TO THE BIRTH OF THE UNIVERSE: *STEPHEN HAWKING'S UNIVERSE: THE COSMOS EXPLAINED*, PUT TOGETHER BY DAVID FILKIN, BECAME AN INTERNATIONALLY AIRED TELEVISION SERIES, AS WELL AS A BOOK, AND HAS FINALLY GIVEN ALL OF US ACCESSIBLE ANSWERS TO:

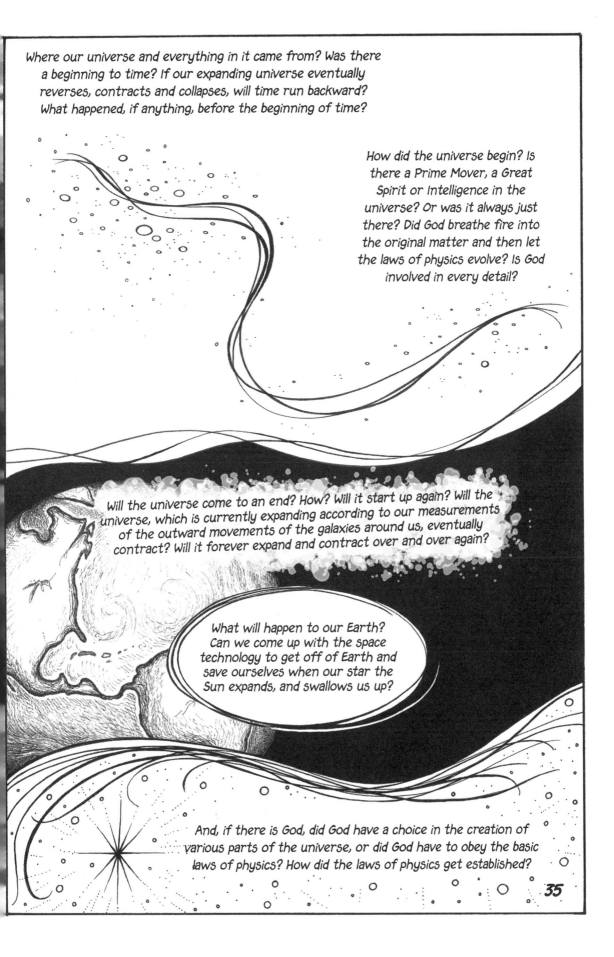

Where our universe and everything in it came from? Was there a beginning to time? If our expanding universe eventually reverses, contracts and collapses, will time run backward? What happened, if anything, before the beginning of time?

How did the universe begin? Is there a Prime Mover, a Great Spirit or Intelligence in the universe? Or was it always just there? Did God breathe fire into the original matter and then let the laws of physics evolve? Is God involved in every detail?

Will the universe come to an end? How? Will it start up again? Will the universe, which is currently expanding according to our measurements of the outward movements of the galaxies around us, eventually contract? Will it forever expand and contract over and over again?

What will happen to our Earth? Can we come up with the space technology to get off of Earth and save ourselves when our star the Sun expands, and swallows us up?

And, if there is God, did God have a choice in the creation of various parts of the universe, or did God have to obey the basic laws of physics? How did the laws of physics get established?

35

PERSONALLY, I ALWAYS THOUGHT IT STRANGE THAT AS I WENT ON IN SCHOOL. THESE QUESTIONS WERE ALMOST NEVER RAISED, THAT FROM OUR MIDTEENS ON, THEY TAUGHT US MOSTLY SPECIALIZED BITS AND PIECES OF INFORMATION SO WE COULD PASS EXAMS AND EARN A LIVING.

BUT WE WERE NOT TAUGHT TO PAY ATTENTION TO EVERYTHING IN THE WORLD AROUND US, AND SO WERE NOT MADE AWARE OF HOW OUR OWN MINDS AND BODIES WORK, TO THEIR CONNECTION, OR HOW IRRATIONALLY HUMANS LIVE THEIR LIVES IN THE FACE OF HOW THE UNIVERSE WORKS. CHILDREN ASK THESE QUESTIONS WHEN THEY ARE YOUNG—AND THEN THEIR QUESTIONS FADE OUT AND THEIR CONNECTION WITH THE NATURAL WORLD IS LOST UNDER THE WEIGHT OF EXAMS AND SOCIETY'S EXPECTATIONS, UNTIL IN ADULTHOOD IT SEEMS OUR BEHAVIOR IS TOTALLY OUT OF HARMONY WITH EVERYTHING ELSE IN NATURE. I'M WITH STEPHEN HAWKING—WE NEED TO UNDERSTAND THE UNIVERSE. OR IT WILL KILL US.

THEORETICAL PHYSICIST STEPHEN HAWKING BROUGHT TOGETHER THE TWO GREAT AND CONTRADICTORY THEORIES OF THE 20TH CENTURY:

EINSTEIN'S THEORY OF RELATIVITY ABOUT THE LARGE, VISIBLE UNIVERSE THAT EXPLAINED THE FORCE OF GRAVITY THAT PULLED MATTER TOGETHER INTO STARS AND GALAXIES, AND PREDICTED THE EXPANSION OF THE UNIVERSE.

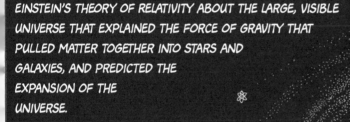

QUANTUM OR PARTICLE PHYSICS THAT EXPLAINS THE ATOM AND THE SUBATOMIC WORLD OF PARTICLES OF MATTER NOT VISIBLE TO THE NAKED EYE.

HAWKING DESCRIBES THE ENTIRE UNIVERSE, MACRO AND MICRO, IN HIS BOOK *A BRIEF HISTORY OF TIME FROM THE BIG BANG TO BLACK HOLES* (NO MATH EQUATIONS SO EVEN PEOPLE LIKE ME CAN UNDERSTAND)

1. ITS ORIGIN

2. ITS FATE

3. HOW IT WORKS— BOTH LARGE, VISIBLE OBJECTS LIKE STARS AND THE SUBATOMIC WORLD OF QUARKS AND QUANTA

I GOT A FEW QUESTIONS, LIKE...

WHEN'S LUNCH!

Where did our universe come from? What happened before time began?

How did the universe begin? Is there a Great Spirit or Intelligence? If so, did God just start things off? Or is She involved in every detail? Does God have to obey the laws of physics like everything else?

Will it all keep expanding forever? Or expand and contract over and over?

WHY ISN'T EVERYONE UP NIGHTS ASKING THESE QUESTIONS?

When our star the Sun expands and swallows the Earth, will we still be here? Or have the technology to get off Earth and find another star?

ALL OVER THE UNIVERSE, THERE ARE
MILLIONS OF CLUSTERS OF GALAXIES FULL
OF BILLIONS OF STARS.

A GALAXY IS COMPOSED OF MILLIONS OF
STAR SYSTEMS LIKE OUR OWN MILKY WAY.
THE MILKY WAY IS ONLY ONE GALAXY IN
THIS LOCAL CLUSTER OF GALAXIES.

OUR STAR, THE SUN, WITH ITS PLANETS, IS
NEAR THE EDGE OF OUR MILKY WAY GALAXY.

# OUR TWO BASIC PHYSICS THEORIES:

## 1. LARGE, VISIBLE UNIVERSE: GENERAL PHYSICS

ALBERT EINSTEIN POSSESSED A RAGE FOR ORDER AND CERTAINTY, AND A CONVICTION THAT "GOD DID NOT PLAY DICE." HE WANTED TO FIND A UNIFIED THEORY OF ALL PHYSICAL LAWS, BUT GOT AS FAR AS THE THEORY OF RELATIVITY.

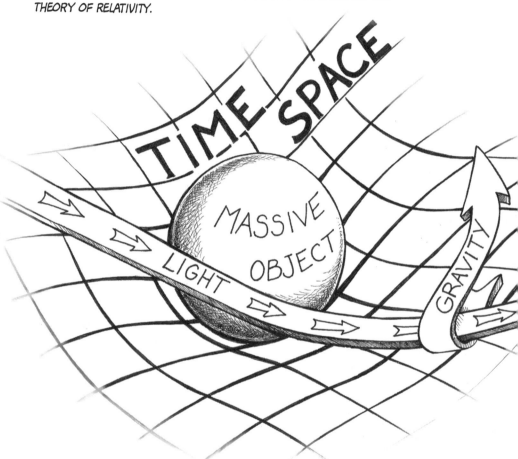

Space and time are not separate. They form a 4-dimensional continuum: height, length, depth, space—called space-time.

Energy and matter are convertible. Mass is a form of energy. The relation between the two is given by the famous equation $E=mc2$ in which c is the speed of light—this is the special relativity theory.

Nothing may travel faster than the speed of light. This is because energy and mass are equal. The energy which an object has due to is motion will increase its mass. So, as an object approaches the speed of light, its mass becomes infinite. The mass becomes so great, it can never reach the speed of light.

So—only light, or other waves with no intrinsic mass, can move at the speed of light.

This General Theory of Relativity also includes The Special Theory that includes gravity—the mutual attraction of all massive bodies.

The force of gravity has the effect of curving space and time. Whenever there is a massive object, like a star or a black hole, the space around it is curved—and as space can never be separated from time in relativity theory, time is also affected by the presence of matter, flowing at different rates in different parts of the universe.

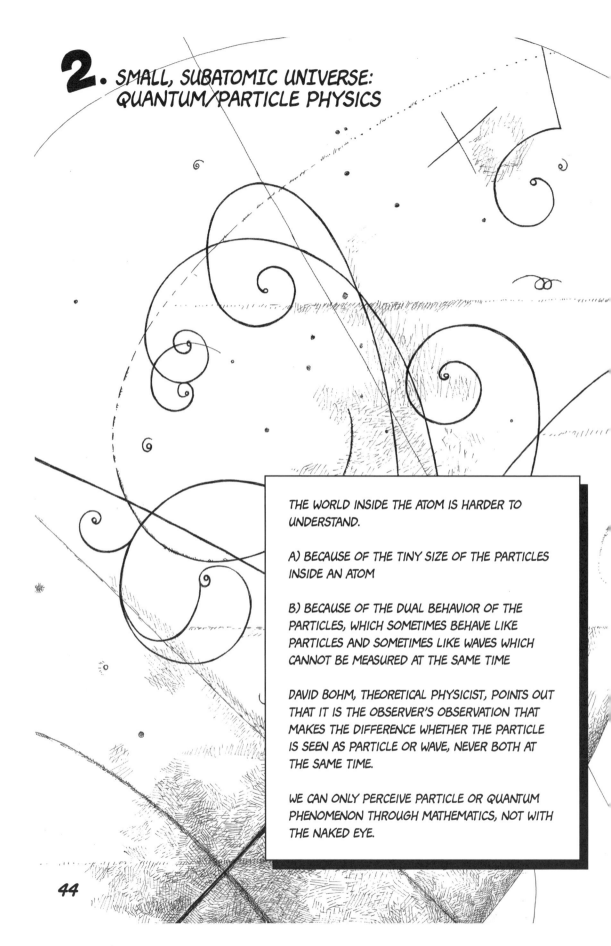

# 2. SMALL, SUBATOMIC UNIVERSE: QUANTUM/PARTICLE PHYSICS

THE WORLD INSIDE THE ATOM IS HARDER TO UNDERSTAND.

A) BECAUSE OF THE TINY SIZE OF THE PARTICLES INSIDE AN ATOM

B) BECAUSE OF THE DUAL BEHAVIOR OF THE PARTICLES, WHICH SOMETIMES BEHAVE LIKE PARTICLES AND SOMETIMES LIKE WAVES WHICH CANNOT BE MEASURED AT THE SAME TIME

DAVID BOHM, THEORETICAL PHYSICIST, POINTS OUT THAT IT IS THE OBSERVER'S OBSERVATION THAT MAKES THE DIFFERENCE WHETHER THE PARTICLE IS SEEN AS PARTICLE OR WAVE, NEVER BOTH AT THE SAME TIME.

WE CAN ONLY PERCEIVE PARTICLE OR QUANTUM PHENOMENON THROUGH MATHEMATICS, NOT WITH THE NAKED EYE.

# LIGHT CAN TAKE THE FORM OF WAVES OR PARTICLES

WHEN MAX PLANCK DISCOVERED THAT THE ENERGY OF HEAT RADIATION APPEARS IN THE FORM OF 'ENERGY PACKETS', EINSTEIN CALLED THEM 'QUANTA'. LIGHT QUANTA GAVE QUANTUM THEORY ITS NAME.

## FOUR FORCES IN THE UNIVERSE

```
1. gravity
2. electromagnetic
3. weak nuclear force
4. strong nuclear force
```

### UNIFIED THEORY
PHYSICISTS ARE STILL LOOKING FOR A UNIFIED THEORY THAT WILL EXPLAIN ALL FOUR FORCES AS DIFFERENT ASPECTS OF A SINGLE FORCE.

WE NOW KNOW THAT EVERY PARTICLE HAS AN ANTI-PARTICLE, A PARTICLE IDENTICAL TO ANOTHER ELEMENTARY PARTICLE IN MASS BUT OPPOSITE TO IT IN ELECTRIC AND MAGNETIC PROPERTIES. A COLLISION ANNIHILATES THEM BOTH IN A FLASH OF LIGHT.

THERE COULD BE WHOLE ANTIWORLDS OF ANTIPEOPLE OUT THERE. WHOLE ANTIWORLDS OF ANTIPARENTS AND ANTITEACHERS AND ANTICATS...

WATCHIT THERE!

So if quantum theory has demolished the classical concepts of solid objects,

we ourselves consist of atoms and the atoms in anything consist mostly of empty space and particles that can behave like waves...

...why can't we walk through closed doors?

Well, there's still the physics of mass to deal with, I think.

What about whether a leaf is really green? Because that is how the human eye and brain interpret it? Remove our human mental person, our human consciousness, and reality changes, doesn't it!

I mean, the leaf isn't really green at all except for the way a human eye sees light and color.

Amazing to think about.

I think that's what some of the great religious philosophers like Krishnamurti meant by meditation. Like, if you sit really quietly and let thoughts and personal stuff die away, something else happens. A silence. I've been reading that all sorts of new insights can happen in silence that aren't twisted up by thought.

*If you young people stay aware like this, according to Darwin's principle of natural selection, there will be variations in the genetic material. This psychological revolution may produce beings rational enough to use more than 15% of our brains which is all we use now.*

*Beings rational enough, free enough of images and conditioning, to observe the universe as it is, and not from positions that cloud the truth.*

**47**

SCIENTISTS INTERESTED IN THE
TRUTH ALWAYS DO EXPERIMENTS
TO SEE IF WHAT THEY THINK
ACTUALLY WORKS.

## EXPERIMENT:

SIT CROSSED-LEGGED OR NORMALLY ON A CHAIR, BACK STRAIGHT SO
LUNGS ARE FREE TO BREATHE, EYES EITHER CLOSED OR FOCUSSED ON
THE FLOOR ABOUT 4 FEET AWAY. DON'T MOVE, BREATHE DEEPLY A FEW
TIMES. THEN—JUST SIT, FOR AT LEAST FIVE MINUTES. SEE WHAT HAPPENS.

IT'S A GOOD WAY TO FIND OUT WHAT'S ON YOUR MIND. YOU CAN BRING
A PROBLEM TO THIS MEDITATION, SOMETHING YOU'RE AFRAID OF,
SOMETHING OR SOMEONE YOU WANT. BUT NO THOUGHTS, NO FANTASIES.
IF THEY OCCUR, BRING YOUR MIND BACK TO JUST SITTING. SEE WHAT NEW
INSIGHTS OCCUR.

# CHAPTER 4
# The Universe Gave Birth to the Earth, and a Creature that Can't Mind Its Own Business

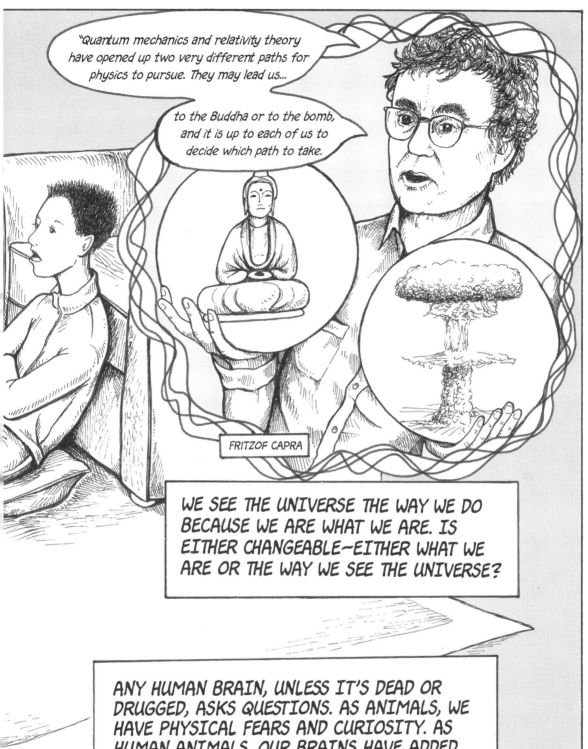

"Quantum mechanics and relativity theory have opened up two very different paths for physics to pursue. They may lead us...

to the Buddha or to the bomb, and it is up to each of us to decide which path to take.

FRITZOF CAPRA

WE SEE THE UNIVERSE THE WAY WE DO BECAUSE WE ARE WHAT WE ARE. IS EITHER CHANGEABLE—EITHER WHAT WE ARE OR THE WAY WE SEE THE UNIVERSE?

ANY HUMAN BRAIN, UNLESS IT'S DEAD OR DRUGGED, ASKS QUESTIONS. AS ANIMALS, WE HAVE PHYSICAL FEARS AND CURIOSITY. AS HUMAN ANIMALS, OUR BRAINS HAVE ADDED PSYCHOLOGICAL FEARS AND CURIOSITY. IF YOU DON'T KNOW WHAT YOUR QUESTIONS ARE, SIT ABSOLUTELY STILL FOR FIVE MINUTES AND ASK YOUR MIND WHAT IT THINKS.

# CHARACTERS IN ORDER OF APPEARANCE:

Births of
The Universe

Galaxies
in General

Stars

Milky Way Galaxy
in Particular

Our Sun in
Particular

Planets

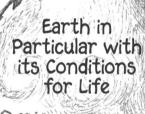

Earth in
Particular with
its Conditions
for Life

# THE UNIVERSE

Quasars - a quasi-stellar object, that may be the nucleus of a galaxy containing a black hole.

Protogalaxies - condensing gas clouds on their way to becoming galaxies.

Dark clouds of dust and gas condensing to form protogalaxies.

Elliptical galaxies full of old stars.

Spiral galaxies full of young stars, like our own Milky Way.

High-energy gamma radiation

Low-energy microwave radiation

Dark matter which we cannot see, only detectable by its gravitational effect on matter we can see.

# GALAXIES

A GALAXY IS SIMPLY A GREAT BUNCH OF STARS HELD TOGETHER BY GRAVITY. MIXED IN WITH THE STARS, WITH THEIR PLANETS AND MOONS, ARE ASTEROIDS (LUMPS OF ROCK), GASES, NEBULAE, INTERSTELLAR MATERIAL. QUASARS, ALTHOUGH THE BRIGHTEST, ARE ABOUT 15 BILLION LIGHT-YEARS AWAY. ANDROMEDA, OUR CLOSEST AND SISTER GALAXY IS ONLY ABOUT FIFTEEN MILLION LIGHT YEARS AWAY. MILKY WAY AND ANDROMEDA BELONG TO OUR LOCAL GROUP OF 19 GALAXIES.

WELL, AT LEAST NOW I KNOW WHERE I AM.

THE MILKY WAY IS A SPIRAL GALAXY. IT HAS A CENTRAL BULGE ENCIRCLED WITH 4 ARMS. WE ARE 2/3 OF THE WAY FROM THE CENTER IN ONE OF THE ARMS. OUR OLDEST STARS, THE AGE OF THE GALAXY ITSELF, MAY BE ABOUT 13 OR 14 BILLION YEARS OLD. THE MILKY WAY IS ABOUT 100,000 LIGHT YEARS ACROSS (A LIGHT YEAR IS ABOUT 5,879 BILLION MILES). OUR SOLAR SYSTEM IS 12 LIGHT-HOURS ACROSS (ABOUT 8 BILLION MILES).

# STARS: BIRTH & DEATH

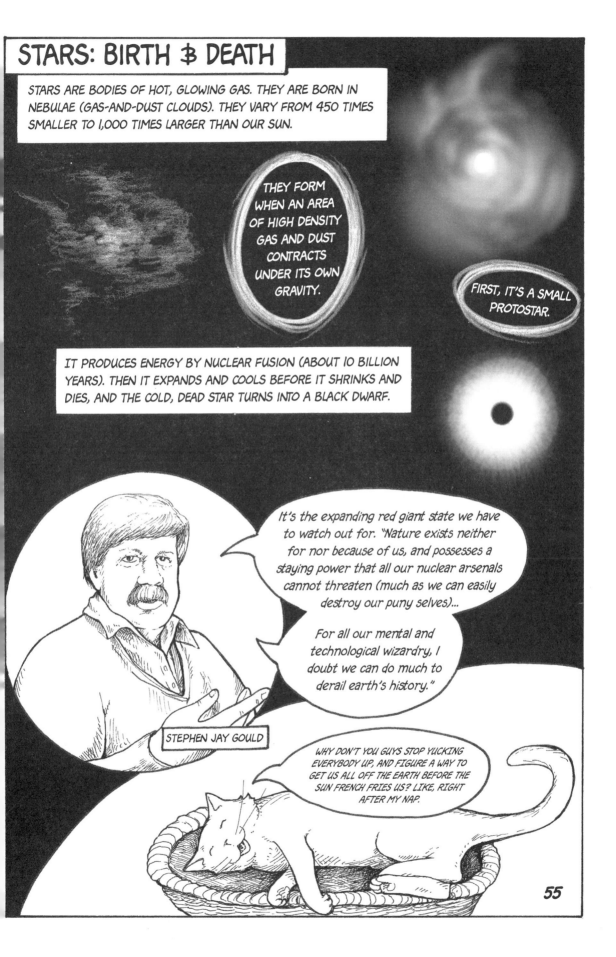

STARS ARE BODIES OF HOT, GLOWING GAS. THEY ARE BORN IN NEBULAE (GAS-AND-DUST CLOUDS). THEY VARY FROM 450 TIMES SMALLER TO 1,000 TIMES LARGER THAN OUR SUN.

THEY FORM WHEN AN AREA OF HIGH DENSITY GAS AND DUST CONTRACTS UNDER ITS OWN GRAVITY.

FIRST, IT'S A SMALL PROTOSTAR.

IT PRODUCES ENERGY BY NUCLEAR FUSION (ABOUT 10 BILLION YEARS). THEN IT EXPANDS AND COOLS BEFORE IT SHRINKS AND DIES, AND THE COLD, DEAD STAR TURNS INTO A BLACK DWARF.

It's the expanding red giant state we have to watch out for. "Nature exists neither for nor because of us, and possesses a staying power that all our nuclear arsenals cannot threaten (much as we can easily destroy our puny selves)...

For all our mental and technological wizardry, I doubt we can do much to derail earth's history."

STEPHEN JAY GOULD

WHY DON'T YOU GUYS STOP YUCKING EVERYBODY UP, AND FIGURE A WAY TO GET US ALL OFF THE EARTH BEFORE THE SUN FRENCH FRIES US? LIKE, RIGHT AFTER MY NAP.

Sun

OUR SOLAR SYSTEM CONSISTS OF OUR STAR (THE SUN), OUR EIGHT PLANETS PLUS THEIR 61 KNOWN MOONS, ASTEROIDS, COMETS, METEOROIDS, GAS, DUST ALL CIRCLING AROUND IT.

Jupiter

Earth

Mercury

Mars

THE SUN IS AN ORDINARY STAR, ABOUT HALFWAY THOUGH ITS LIFE, BORN ALONG WITH EARTH AND THE OTHER PLANETS AROUND 5 BILLION YEARS GO.

Venus

Saturn

IT WILL COOL AND BECOME EXTINCT IN ABOUT 4 BILLION YEARS.

Neptune

Uranus

ARGHHH.

## THE BIRTH OF EARTH & OUR COMPANION PLANETS AND MOON

SCIENTISTS THINK PARTICLES OF DUST AND ROCK SWIRLING AROUND THE SUN COLLIDED AND CLUMPED TOGETHER TO FORM THE PLANETS.

EARTH REMAINED ROILING HOT INSIDE BUT COOLED ON THE SURFACE TO A ROCKY CRUST. DEEP INSIDE THE EARTH HOT TEMPERATURES STILL KEEP ROCK MOLTEN. 2 1/2 TO 6 MILLION YEARS AGO, THE CRUST SPLIT UP INTO MOVING PLATES THAT EVENTUALLY TURNED INTO CONTINENTS.

GASES ESCAPING FROM VOLCANOES PROVIDED THE BEGINNINGS OF THE FIRST PERMANENT ATMOSPHERE. OCEANS WERE FORMED FROM EARTH'S ESCAPING VAPORS, ACCUMULATING AND FALLING AS RAIN.

JUST LIKE YOU ORBIT AROUND ME.

OUR MOON IS THE SAME AGE AS OUR EARTH–ABOUT 4.5 BILLION YEARS OLD. THEY ARE LOCKED TOGETHER BY GRAVITY IN THEIR JOINT GRAVITY–HELD IN ORBIT AROUND THE SUN.

WHO MADE YOU THE CENTER OF EVERYTHING?

AND DON'T TELL ME THE CAT GOD.

# LIFE

**WATER & ATMOSPHERE**
ABOUT 70% OF THE EARTH'S SURFACE IS COVERED WITH WATER.

EARTH'S ATMOSPHERE SCREENS OUT SOME OF THE SUN'S HARMFUL RADIATION AND TRAPS ENOUGH HEAT TO PREVENT EXTREMES OF COLD. WATER AND ATMOSPHERE SEEM NECESSARY FOR THE PRESENCE OF LIFE AS WE KNOW IT.

BUT LET NO ONE—SCIENTISTS OR RELIGIOUS LEADERS—TELL YOU WE KNOW EXACTLY HOW LIFE BEGAN. WE DON'T.

HOW DOES WATER, PLUS A FEW AMINO ACIDS AND OTHER SUBSTANCES, JUST TURN INTO LIFE AFTER A FEW MILLION YEARS? MOLECULAR BIOLOGY MAY DO FOR EVOLUTIONARY BIOLOGY WHAT QUANTUM PHYSICS DID FOR COSMOLOGY AND ASTROPHYSICS.

UNDERSTANDING THE VERY SMALL CHANGES OUR UNDERSTANDING OF HOW LARGER BODIES WORK.

I KEEP TELLING MY MOTHER THE SAME THING.

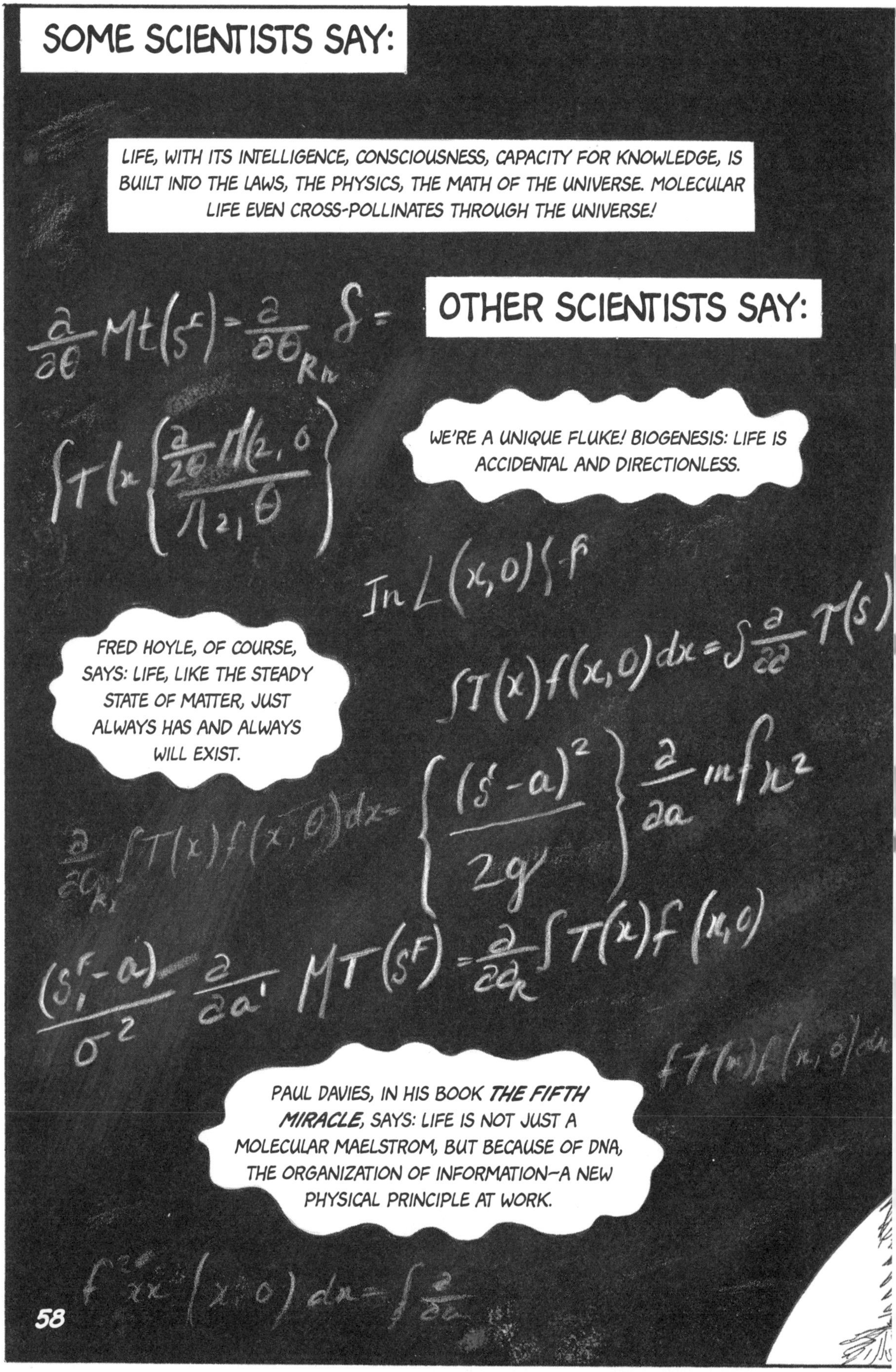

SOME SCIENTISTS SAY:
LIFE, WITH ITS INTELLIGENCE, CONSCIOUSNESS, CAPACITY FOR KNOWLEDGE, IS BUILT INTO THE LAWS, THE PHYSICS, THE MATH OF THE UNIVERSE. MOLECULAR LIFE EVEN CROSS-POLLINATES THROUGH THE UNIVERSE!
OTHER SCIENTISTS SAY:
WE'RE A UNIQUE FLUKE! BIOGENESIS: LIFE IS ACCIDENTAL AND DIRECTIONLESS.
FRED HOYLE, OF COURSE, SAYS: LIFE, LIKE THE STEADY STATE OF MATTER, JUST ALWAYS HAS AND ALWAYS WILL EXIST.
PAUL DAVIES, IN HIS BOOK ***THE FIFTH MIRACLE***, SAYS: LIFE IS NOT JUST A MOLECULAR MAELSTROM, BUT BECAUSE OF DNA, THE ORGANIZATION OF INFORMATION—A NEW PHYSICAL PRINCIPLE AT WORK.

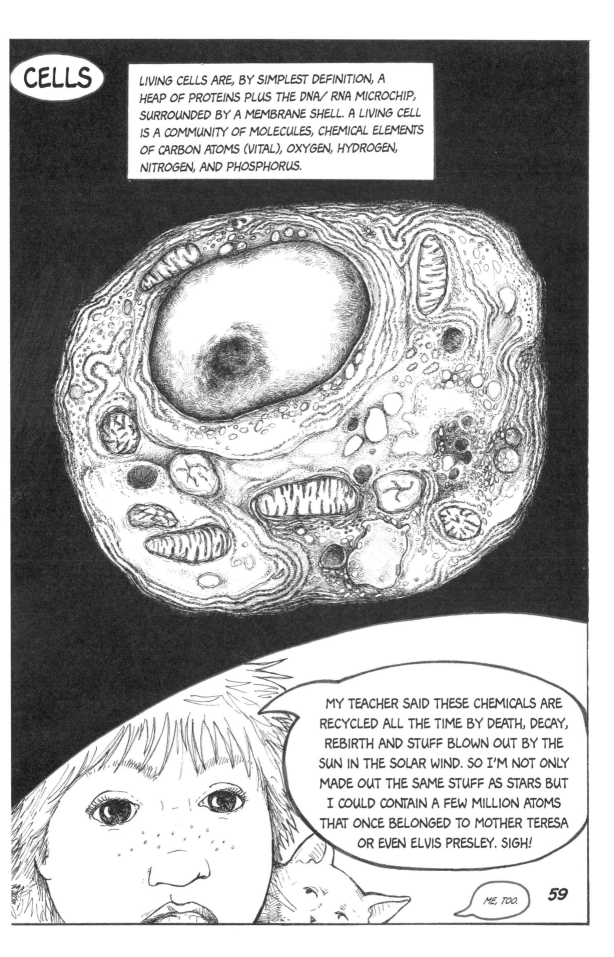

**4.6 bya:** oldest rocks

**3.8 bya:** first evidence of life, single-celled bacteria— and DNA, which distinguishes life from just molecules

**2.5 bya:** first remains of multicelled animals, worm-like organisms

**530 mya:** first spine-like structure in a tiny eel-like animal, ancestor of vertebrates like us

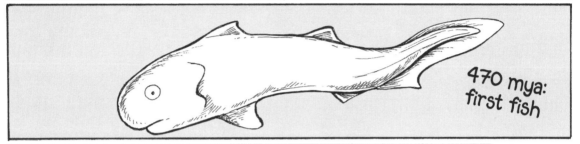

**470 mya:** first fish

**450 mya: first land plants and animals**

NOT ME YET!

**360 mya:** four-legged animals with backbones

**325 mya: first forests**

**310 mya:** reptiles—we are independent of water with egg-laying reptiles

**248 mya: 96 percent of all species of animals and plants wiped out!**

**215 mya:** first true mammals, which gave birth to live young—tiny, furry shrew-like creature

**150 mya:** first birds and feathered dinosaurs

**65 mya:** asteroid hits Earth, changes climate, wipes out 76 percent of all species, including dinosaurs

**53 mya:** first whales

**20 mya:** grasses and horses

**5 mya:** world's climate began cooling

around **2 mya** first glaciers formed

around **10,000 ya** ice had retreated— gone forever? coming back?

around **500 ya** first type of modern humans similar to us appeared in Africa

around **40 ya**, we were spread around the world

THAT'S ME.

*From the Big Bang on, we were off and running!*

# PART TWO
## Where Are We Now?
### Your Brain, Its Body, and Where They Live

# CHAPTER 5
# YOUR BRAIN AND ITS BODY

*HUMAN BEINGS, TEENAGERS, AND
THE INEXORABLE LAWS OF NATURE*

IT'S A KNOWN FACT THAT HUMAN BEINGS CANNOT ULTIMATELY
BREAK THE FUNDAMENTAL LAWS OF THE UNIVERSE. ADULTS
DEALING WITH TEENAGERS SOMETIMES FORGET THIS.

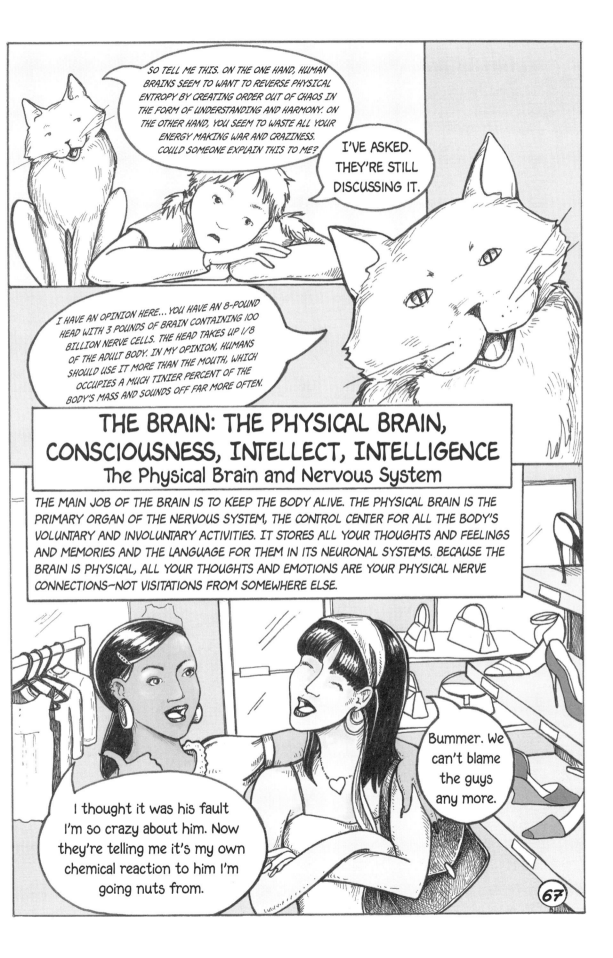

SO TELL ME THIS. ON THE ONE HAND, HUMAN BRAINS SEEM TO WANT TO REVERSE PHYSICAL ENTROPY BY CREATING ORDER OUT OF CHAOS IN THE FORM OF UNDERSTANDING AND HARMONY. ON THE OTHER HAND, YOU SEEM TO WASTE ALL YOUR ENERGY MAKING WAR AND CRAZINESS. COULD SOMEONE EXPLAIN THIS TO ME?

I'VE ASKED. THEY'RE STILL DISCUSSING IT.

I HAVE AN OPINION HERE... YOU HAVE AN 8-POUND HEAD WITH 3 POUNDS OF BRAIN CONTAINING 100 BILLION NERVE CELLS. THE HEAD TAKES UP 1/8 OF THE ADULT BODY. IN MY OPINION, HUMANS SHOULD USE IT MORE THAN THE MOUTH, WHICH OCCUPIES A MUCH TINIER PERCENT OF THE BODY'S MASS AND SOUNDS OFF FAR MORE OFTEN.

# THE BRAIN: THE PHYSICAL BRAIN, CONSCIOUSNESS, INTELLECT, INTELLIGENCE
## The Physical Brain and Nervous System

THE MAIN JOB OF THE BRAIN IS TO KEEP THE BODY ALIVE. THE PHYSICAL BRAIN IS THE PRIMARY ORGAN OF THE NERVOUS SYSTEM, THE CONTROL CENTER FOR ALL THE BODY'S VOLUNTARY AND INVOLUNTARY ACTIVITIES. IT STORES ALL YOUR THOUGHTS AND FEELINGS AND MEMORIES AND THE LANGUAGE FOR THEM IN ITS NEURONAL SYSTEMS. BECAUSE THE BRAIN IS PHYSICAL, ALL YOUR THOUGHTS AND EMOTIONS ARE YOUR PHYSICAL NERVE CONNECTIONS—NOT VISITATIONS FROM SOMEWHERE ELSE.

I thought it was his fault I'm so crazy about him. Now they're telling me it's my own chemical reaction to him I'm going nuts from.

Bummer. We can't blame the guys any more.

67

# INTELLECT AND INTELLIGENCE

THE BRAIN HAS 2 MAIN CAPACITIES, BESIDES KEEPING THE BODY ALIVE.

INTELLECT IS THOUGHT, WHICH IS ALWAYS BASED ON THE PAST. THE BRAIN ACTS LIKE A TAPE RECORDER. WHAT IT RECORDS, WE CALL THOUGHT. SO THOUGHT IS JUST MEMORIES—4 BILLION YEARS OF THEM, FROM YOUR PERSONAL LIFE RIGHT NOW ALL THE WAY BACK TO WHEN WE WERE AMOEBA.

INTELLIGENCE IS NEW INSIGHT, NEW THINKING DIFFERENTLY ABOUT OLD THOUGHTS

But we still need thought, right?

THOUGHT IS A SURVIVAL TOOL, SEPARATING A LION'S ATOMS FROM AN APPLE'S ATOMS SO WE KNOW WHEN TO RUN AND WHEN TO EAT. THOUGHT IS NECESSARY OR WE WOULDN'T REMEMBER WORDS OR WHERE WE LIVE OR THE TECHNOLOGY TO MAKE MACHINES OR MEDICINE.

So, thought invents technology, and insight tells us whether to make bombs or medicine out of it?

YOU GOT IT! INSIGHT IS NECESSARY TO KNOW WHAT IS ETHICALLY APPROPRIATE TO DO WITH THOSE THINGS THOUGHT COMES UP WITH. THOUGHT MUST SERVE INSIGHT, NOT CONTROL IT.

KISHORE KHAIRNAR, PHYSICIST: "AMONG LIVING FORMS, THE HUMAN HAS EVOLVED WITH THE ABILITY TO THINK, TO OBSERVE, TO LEARN, TO PASS ON KNOWLEDGE AND CULTURAL IDEAS. THIS ABILITY HAS BROUGHT INTO EXISTENCE AN ALTOGETHER NEW WORLD, THE INNER PSYCHOLOGICAL WORLD OF HUMAN BEINGS. WE MUST UNDERSTAND OUR PSYCHOLOGY SO THAT WE SUFFER AND MAKE OTHERS SUFFER LESS."

# CONSCIOUSNESS IS THE AWARENESS OF BEING AWARE

WE DON'T KNOW WHERE THE MYSTERY OF CONSCIOUSNESS COMES FROM—ONLY THAT IT SEEMS ALL LIFE IS AWARE AND WHAT WE CALL THE HUMAN SPECIES SEEMS ALONE AWARE OF BEING AWARE. WE DON'T KNOW WHERE IN THE HUMAN BRAIN CONSCIOUSNESS RESIDES, BUT MOST NEUROSCIENTISTS AGREE IT TAKES THE STUFF IN HUMAN BRAINS TO BE SELF-CONSCIOUS.

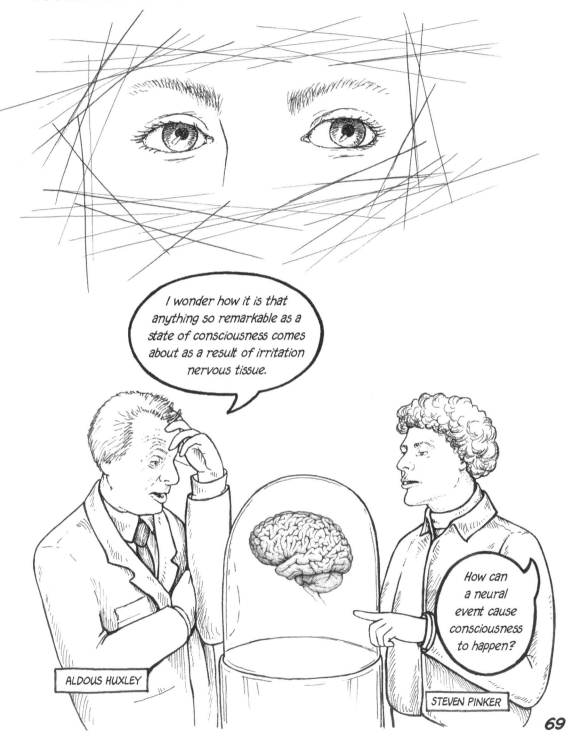

# The Brain

## 3 VISIBLE BRAIN AREAS

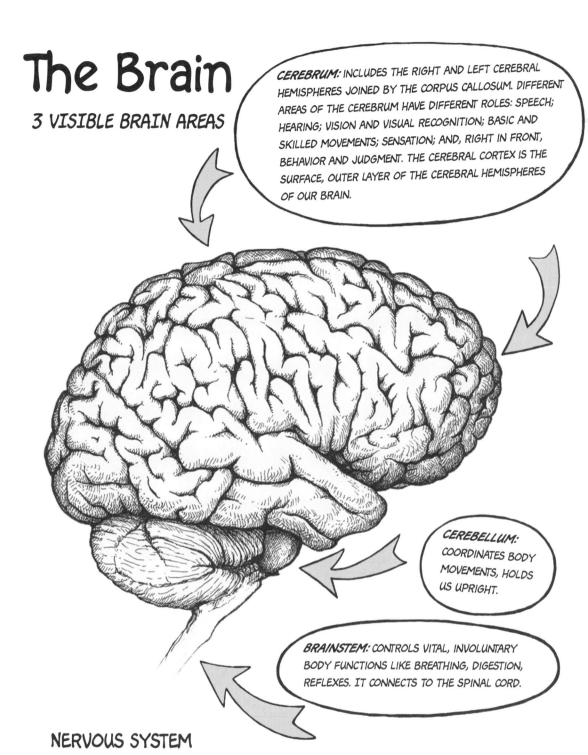

**CEREBRUM:** INCLUDES THE RIGHT AND LEFT CEREBRAL HEMISPHERES JOINED BY THE CORPUS CALLOSUM. DIFFERENT AREAS OF THE CEREBRUM HAVE DIFFERENT ROLES: SPEECH; HEARING; VISION AND VISUAL RECOGNITION; BASIC AND SKILLED MOVEMENTS; SENSATION; AND, RIGHT IN FRONT, BEHAVIOR AND JUDGMENT. THE CEREBRAL CORTEX IS THE SURFACE, OUTER LAYER OF THE CEREBRAL HEMISPHERES OF OUR BRAIN.

**CEREBELLUM:** COORDINATES BODY MOVEMENTS, HOLDS US UPRIGHT.

**BRAINSTEM:** CONTROLS VITAL, INVOLUNTARY BODY FUNCTIONS LIKE BREATHING, DIGESTION, REFLEXES. IT CONNECTS TO THE SPINAL CORD.

## NERVOUS SYSTEM

BRAIN, SPINAL CORD, AND NERVES THROUGHOUT THE BODY. THIS IS THE BODY'S ELECTROCHEMICAL COMMUNICATIONS SYSTEM, SENDING CHEMICAL SIGNALS AS ELECTRICAL IMPULSES FROM THE BRAIN THROUGH THE BODY.

THE BRAIN RECEIVES INFORMATION
1. EXTERNALLY THROUGH THE FIVE SENSES (EYES, EARS, NOSE, SKIN, PALATE)
2. INTERNALLY THROUGH MEMORY-IMAGE-RECOGNITION GENETIC-CULTURAL-AGENDA

# THE SELF—NOBODY'S HOME

REMEMBER THERE IS NO 'I' OR 'ME' OR 'SELF' IN THERE, JUST BUNCHES OF BRAIN NERVE CELLS THAT INVENT THE 'GREAT AND DIFFERENT I AM' OUT OF YOUR STORIES AND MEMORIES TO GIVE YOUR BODY A SENSE OF SECURITY. IT'S A TRICK OF THE BRAIN: IT FEELS AS IF THERE IS A 'ME' EVEN THOUGH THERE ISN'T. AS A SPECIES, WE'RE AS ALIKE AS SNOWFLAKES.

DANIEL DENNETT, IN HIS BOOK *CONSCIOUSNESS EXPLAINED*, SAYS, "THE TROUBLE WITH BRAINS IS WHEN YOU LOOK IN THEM, YOU DISCOVER THAT THERE'S NOBODY HOME. NO PART OF THE BRAIN IS THE THINKER THAT DOES THE THINKING OR THE FEELER THAT DOES THE FEELING." "A SELF," HE SAYS, "IS JUST A COLLECTION OF BIOGRAPHICAL STORIES WE TELL OURSELVES TO GIVE US CONTINUITY IN THIS LIFE, AND, HOPEFULLY, IMMORTALITY IN THE NEXT. SO INSTEAD OF DYING WE CAN GO ON AND ON AND ON AND…"

What a relief! People keep telling me to "be myself" or "find myself" or be "true to myself." I'm nobody. I can stop looking.

I just hear a lot of voices, usually arguing, inside my head. I can give up now worrying about who is the real me. Nobody!

I COULD HAVE TOLD YOU THAT.

71

# MENTAL ILLNESS

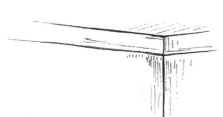

HANNAH CARLSON, BEHAVIORAL SPECIALIST, SAYS IN HER BOOK *COURAGE TO LEAD: START YOUR OWN SUPPORT GROUP FOR MENTAL ILLNESSES AND ADDICTIONS*: "MENTAL ILLNESS INVOLVES DISORDERS OF THINKING, FEELING, JUDGMENT, AND BEHAVIOR...MENTAL ILLNESS SHOULD BE THOUGHT OF LIKE ANY OTHER PHYSICAL ILLNESS AND DIAGNOSED BY ITS SYMPTOMS."

Many scientists, psychiatrists, and psychologists cite the causes of mental illness as particular genes, chemicals in the brain, or social and environmental influences—generally a combination of physiological, neurochemical, psychological, and social or environmental causes. Nerve cell chemicals, called neurotransmitters, such as dopamine, norephinephrin, and serotonin, have been linked to mental illnesses such schizophrenia, bipolar disorder, obsessive-compulsive disorder, depression, ADHD, chemical addictions. Like other physical, medical conditions, mental illnesses can run in families.

DON'T LOOK SO SMUG. THIS INCLUDES THE CAT FAMILY.

DON'T BE SILLY. THERE'S NO ALCOHOLISM OR INSANITY IN MY FAMILY.

**THE HEAD:** THE HEAD HOLDS YOUR EYES, EARS, NOSE, AND MOUTH FOR NOURISHMENT AND COMMUNICATION. IT IS SURROUNDED BY HAIR AND SKIN FOR TOUCHING AND PROTECTING. IT HOLDS THE BRAIN WHICH COORDINATES YOUR LIFE AND CALLS YOUR SHOTS—INCLUDING YOUR SEX LIFE. ITS LOSS COULD AFFECT YOUR LIFE BADLY.

# BODY PARTS

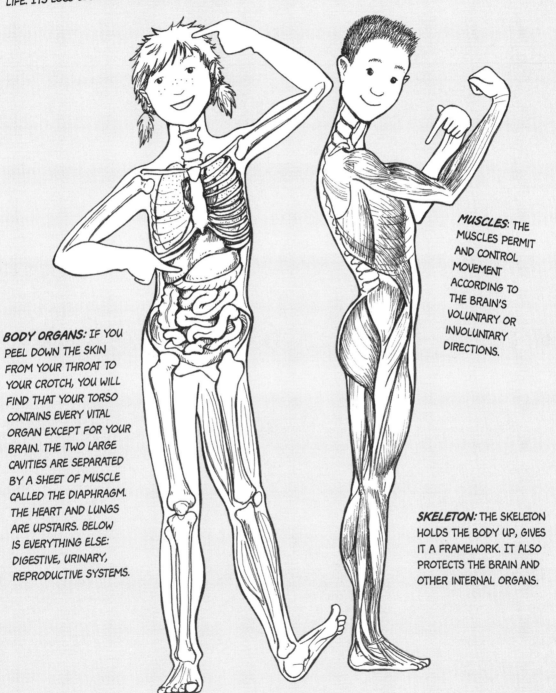

**MUSCLES**: THE MUSCLES PERMIT AND CONTROL MOVEMENT ACCORDING TO THE BRAIN'S VOLUNTARY OR INVOLUNTARY DIRECTIONS.

**BODY ORGANS:** IF YOU PEEL DOWN THE SKIN FROM YOUR THROAT TO YOUR CROTCH, YOU WILL FIND THAT YOUR TORSO CONTAINS EVERY VITAL ORGAN EXCEPT FOR YOUR BRAIN. THE TWO LARGE CAVITIES ARE SEPARATED BY A SHEET OF MUSCLE CALLED THE DIAPHRAGM. THE HEART AND LUNGS ARE UPSTAIRS. BELOW IS EVERYTHING ELSE: DIGESTIVE, URINARY, REPRODUCTIVE SYSTEMS.

**SKELETON:** THE SKELETON HOLDS THE BODY UP, GIVES IT A FRAMEWORK. IT ALSO PROTECTS THE BRAIN AND OTHER INTERNAL ORGANS.

# CHAPTER 6
# GENES, SEX, AND MISTAKES

ANY TEENAGER ALREADY
KNOWS WHAT SCIENTISTS
KEEP TRYING TO PROVE:
SEX IS MINDLESS
AND NOTHING IS PERFECT.

But I
love him,
Mother!

HUMANS MAKE SUCH
A FUSS ABOUT THE
SIMPLEST THINGS.

# IT'S OUR GENES THAT WANT SEX

OUR BODIES WORK PRETTY WELL. OUR WHOLE ECOSYSTEM WORKS PRETTY WELL. THIS IS NOT BECAUSE MOTHER NATURE FORESAW AND PLANNED FOR YOUR PERSONAL PERFECTION.

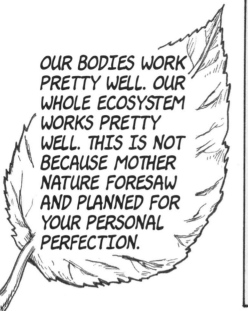

**1** THIS IS BECAUSE, AS DARWIN POINTED OUT, THERE IS A PURELY SELFISH STRUGGLE GOING ON AMONG LIVING CREATURES TO SURVIVE.

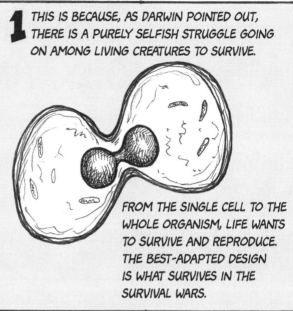

FROM THE SINGLE CELL TO THE WHOLE ORGANISM, LIFE WANTS TO SURVIVE AND REPRODUCE. THE BEST-ADAPTED DESIGN IS WHAT SURVIVES IN THE SURVIVAL WARS.

**2** WE ARE THE RESULT OF PREVIOUS MISTAKES (MUTATIONS) BECAUSE GENES DO NOT ALWAYS REPRODUCE EXACTLY.

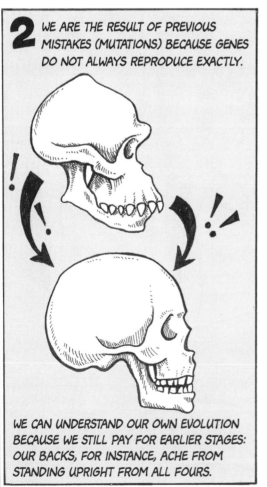

WE CAN UNDERSTAND OUR OWN EVOLUTION BECAUSE WE STILL PAY FOR EARLIER STAGES: OUR BACKS, FOR INSTANCE, ACHE FROM STANDING UPRIGHT FROM ALL FOURS.

TELL ME ABOUT SUPERIOR DESIGN. MY BACK DOESN'T ACHE.

# THE SELFISH GENE

RICHARD DAWKINS, AUTHOR OF *THE SELFISH GENE*, STEPHEN PINKER, STEVEN JAY GOULD, DANIEL DENNETT, OLIVER SACHS: THE STAND-UP COMICS OF THE SCIENCE-WRITERS 'WHAT-IS-A-HUMAN-BEING' WORLD WOULD ALL NOD AND CLAP FOR PINKER WHEN HE SAYS:

"PEOPLE DON'T SELFISHLY SPREAD THEIR GENES. GENES SELFISHLY SPREAD THEMSELVES. THEY DO IT BY THE WAY THEY BUILD OUR BRAINS...BY MAKING US ENJOY LIFE, HEALTH, SEX, FRIENDS, AND CHILDREN."

GENES ARRANGE FOR NERVE/ CHEMICAL PLEASURE WHEN WE HAVE SEX, AND OBEY THEIR OTHER DIRECTIVES FOR REPRODUCTION LIKE EATING HEALTHY FOOD, STAYING ALIVE, STICKING CLOSE TO A SUPPORT SYSTEM, MARRYING INTO OUR OWN GENE POOL INSTEAD OF CHICKENS.

GENES ARE NOT NECESSARILY DESTINY: WE PROVE WE ARE SMARTER THAN OUR GENES WHEN WE USE CONTRACEPTION.

DIFFERENCES IN BRAINS LIKE TALENT OR I.Q. OR THE ABILITY TO JUGGLE OR SPEAK SANSKRIT ARE MINOR CLEVERNESSES FOR IMPRESSING EACH OTHER, FOR ESTABLISHING STATUS, OR AN ATTEMPT AT IMMORTALITY. ALL HUMAN BRAINS ARE BASICALLY ALIKE, 'ORGANS OF COMPUTATION ENGINEERED BY NATURAL SELECTION,' AS PINKER SAYS.

IN HIS GENETICALLY-ENGINEERED THEORY OF BEHAVIOR BASED ON NATURAL SELECTION OF THE FITTEST TO SURVIVE, THERE ARE SOME INTERESTING POINTS.

THIS BETTER BE GOOD. I'M GOING TO SLEEP AGAIN.

AREN'T YOU INTERESTED IN THE QUESTION, "WHAT IS A HUMAN BEING?"

YOU'RE KIDDING, RIGHT?

# THE GREATEST QUESTION OF ALL

SIGH!

THIS BETTER BE GOOD.

WHY DO OUR GENETICALLY-ENGINEERED BRAINS, OUR MENTAL PHYSIOLOGY, LACK THE COMPUTATIONAL SKILLS NECESSARY TO SOLVE THE BIGGEST OF HUMAN PHILOSOPHICAL PROBLEMS?

YOU WANNA REPEAT THOSE QUESTIONS?

CATS HAVE ALL THEIRS LICKED.

WHAT IS INTELLIGENCE? AWARENESS? CONSCIOUSNESS? IS THERE ANY FREE WILL? WHAT, IF ANYTHING, IS GOD? WHERE?

ONLY THE CAT GOD KNOWS.

We are organisms. Our minds are organs, not pipelines to the truth.

We evolved skills to solve life-and-death problems only. We fill, not the niche of the swift, the muscular, the winged, or the fanged, but the niche of the nerds— nerds, not prophetic angels.

STEVEN PINKER

KRISHNAMURTI SAYS, IN HIS BOOK **WHAT ARE YOU DOING WITH YOUR LIFE? BOOKS ON LIVING FOR TEENS**: THERE IS MORE THAN COGNITIVE MACHINERY, INFORMATION-GATHERING INTELLECT, IN OUR BRAINS. WE HAVE ANOTHER CAPACITY—HUMAN AWARENESS, OBSERVATION AND INSIGHT INTO EXPERIENCE, OUR RELATIONSHIP TO EACH OTHER AND THE OUTSIDE WORLD. THIS INSIGHT CAN CAUSE A MUTATION IN OUR BRAINS AND CHANGE OUR BEHAVIOR!

THERE IS NO WAY TO MAKE THE WHOLE BUSINESS OF YOUR DNA SHORT AND FUNNY. BUT A GOOD QUESTION IS, IS THE ORIGIN OF THE GENETIC CODE THE KEY TO THE ORIGIN OF LIFE ITSELF? FIND OUT—YOU'LL GET ALL THE NOBEL PRIZES.

SINCE WE'VE ALREADY BEEN THROUGH 5 MASS EXTINCTIONS, AND WE'RE GOING THROUGH THE 6TH NOW, THE ORIGIN OF DNA, OF LIFE AND REPRODUCTION, MIGHT BE A GOOD SECRET TO TAKE ALONG TO THE NEXT PLACE IN THE UNIVERSE WE CAN FIND A HOME

# THE CELL NUCLEUS: HOME TO YOUR DNA MOLECULE

## HUMAN CELL STRUCTURE

EACH CELL HAS AN OUTER MEMBRANE AND CONTAINS A FLUID MATERIAL (CYTOPLASM) FULL OF SPECIALIZED STRUCTURES (ORGANELLES), ESPECIALLY THE NUCLEUS. MITOCHONDRIA, FOR INSTANCE, ARE LITTLE FILAMENTS WHICH ARE THE SOURCE OF ENERGY IN THE CELL, SYNTHESIZING PROTEIN AND LIPID METABOLISM, AS ANY MOLECULAR BIOLOGIST (OR *TABER'S CYCLOPEDIC MEDICAL DICTIONARY*) WILL TELL YOU.

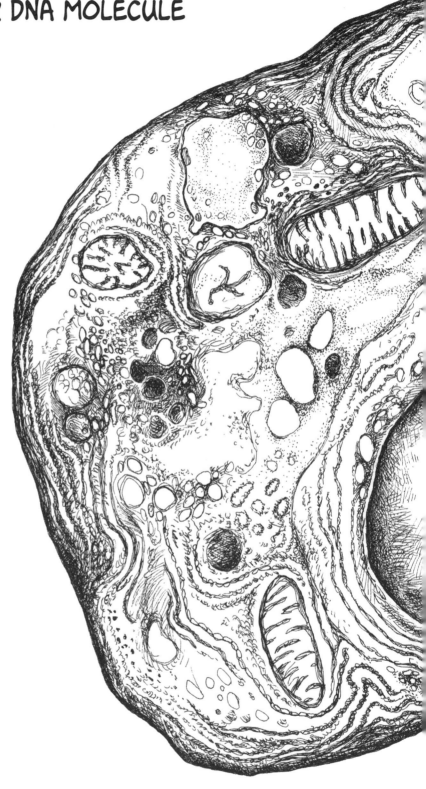

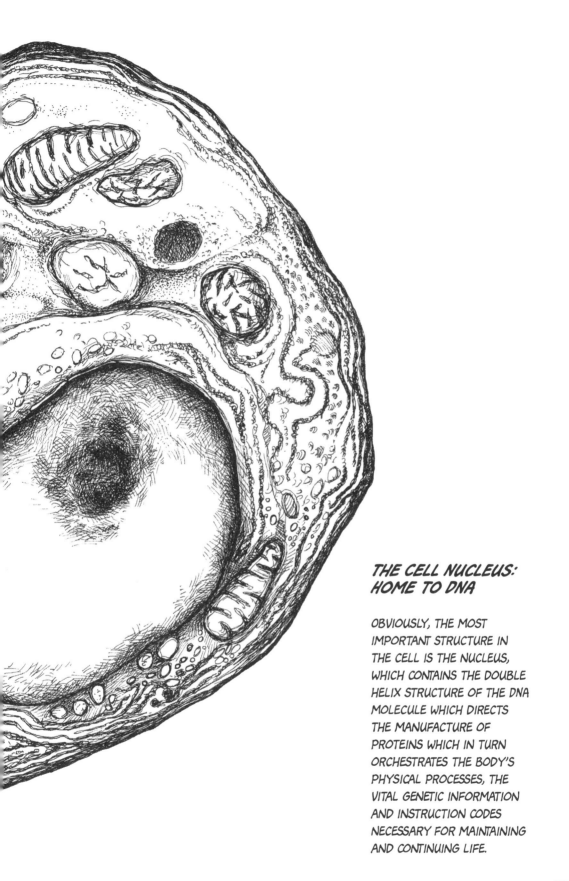

## THE CELL NUCLEUS: HOME TO DNA

OBVIOUSLY, THE MOST IMPORTANT STRUCTURE IN THE CELL IS THE NUCLEUS, WHICH CONTAINS THE DOUBLE HELIX STRUCTURE OF THE DNA MOLECULE WHICH DIRECTS THE MANUFACTURE OF PROTEINS WHICH IN TURN ORCHESTRATES THE BODY'S PHYSICAL PROCESSES, THE VITAL GENETIC INFORMATION AND INSTRUCTION CODES NECESSARY FOR MAINTAINING AND CONTINUING LIFE.

# THE DNA
# MOLECULE

THE GOAL OF THE HUMAN GENOME PROJECT, ACCORDING TO *U.S. NEWS & WORLD REPORT*, IS TO DECIPHER THE HUMAN GENETIC CODE, DETERMINE THE EXACT ORDER OF THE THREE BILLION CHEMICAL LETTERS THAT MAKE UP OUR DNA, THE GENOME WHICH INCLUDES 100,000 OR SO GENES, THE COMPLETE SET OF GENETIC INSTRUCTIONS THAT IS OFTEN CALLED THE BOOK OF LIFE, WITH ITS TWENTY-THREE 'VOLUMES', OR PAIRS OF CHROMOSOMES.

WITHIN THE CELL NUCLEUS, THE DNA IS ORGANIZED INTO FORTY-SIX CHROMOSOMES, TWENTY-THREE PAIRS OF THESE LINEAR THREADS THAT CONTAIN OUR GENES, OUR HEREDITARY DETERMINERS, WITH THE GENES ARRAYED LIKE BEADS ON A NECKLACE. THE DNA MOLECULE CONSISTS OF TWO STRANDS, SO INTERTWINED THEY LOOK LIKE A TWISTED LADDER.

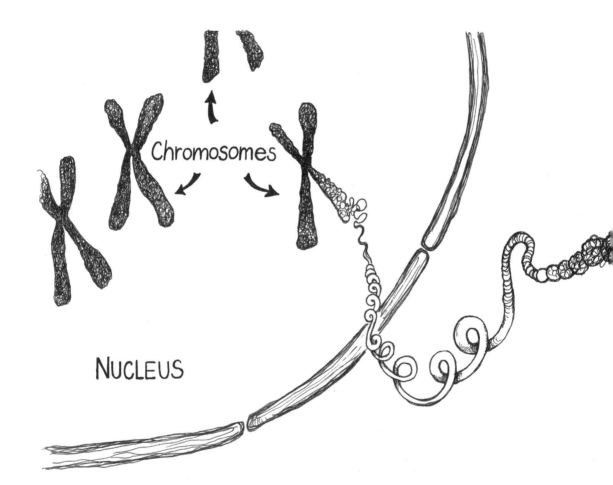

Chromosomes

NUCLEUS

DNA INSTRUCTIONS ARE COPIED ONTO RNA, RNA LEAVES THE CELL NUCLEUS AND ENTERS THE CYTOPLASM, WHERE IT ATTACHES TO A RIBOSOME, A LITTLE PROTEIN FACTORY. THE RIBOSOME READS RNA'S INSTRUCTIONS AND CONNECTS AMINO ACIDS TOGETHER TO BUILD A PROTEIN. THE TWENTY DIFFERENT AMINO ACIDS ARE THE BUILDING BLOCKS OF PROTEINS, WHICH PROVIDE THE STRUCTURAL COMPONENTS OF CELLS AND TISSUES.

REMEMBERING THAT IT IS THE GENE THAT CONTAINS THE RECIPE FOR MAKING A PROTEIN, IT IS THE GENE SEQUENCE, THE INFORMATION BURIED IN THE GENES, THAT SCIENCE USES FOR VITAL INFORMATION: TO TRACE HUMAN ORIGINS (DNA CLUES REVEAL ALL SIX BILLION OF US TRACE OUR ANCESTRY BACK TO A GROUP OF 50,000 AFRICAN HUMANS WHO LIVED 150,00 YEARS AGO; TO SOLVE CRIMES; TO UNCOVER CLUES TO HUMAN DISEASES AND PERSONALIZE MEDICINE ACCORDING TO GENETIC PROFILE.

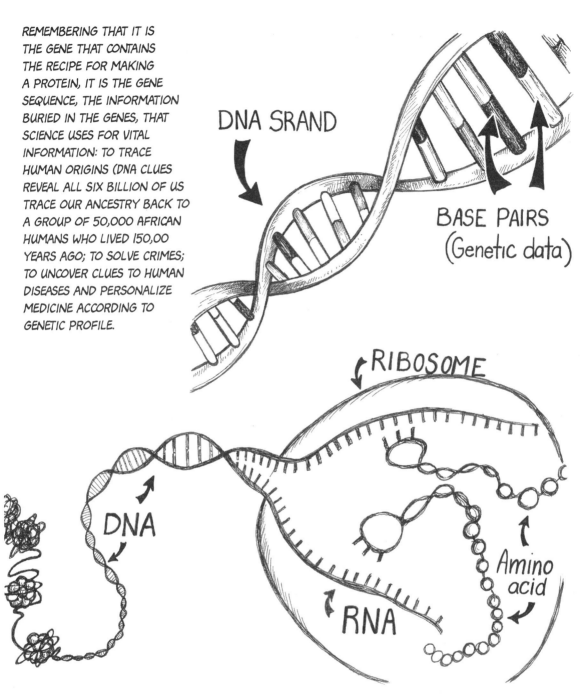

DNA SRAND

BASE PAIRS
(Genetic data)

RIBOSOME

DNA

RNA

Amino acid

## COMPARING GENOMES

FOR THOSE WHO LOVE TO DIFFER INSTEAD OF RELATE, AT THE GENOMIC LEVEL THEY SHOULD BE HAPPY TO KNOW THAT LIKE THEIR FINGERPRINTS, THEIR GENETIC DATA ARE UNIQUE. AT THIS LEVEL OF DNA, EACH PERSON IS UNIQUE, WITH MILLIONS OF DISTINGUISHING GENETIC CHANGES THAT INFLUENCE OUR LOOKS, THE DISEASES WE MIGHT INHERIT, OUR POSSIBLE BEHAVIOR. PAY ATTENTION TO THE CONDITIONAL PHRASING OF THE SCIENTISTS: 'INFLUENCE', 'MIGHT' AND 'POSSIBLE'. IT TAKES MORE THAN A FINGERPRINT, GENETIC OR OTHERWISE, TO DETERMINE MORAL CHARACTER, TALENT, MENTAL HEALTH OR OTHERWISE, THE CAPACITY TO FIND THE MEANING OF LIFE.

GENES ARE NOT ALL
THERE IS TO IT

ALSO, IT IS NECESSARY TO PAY ATTENTION TO AGE AND DEVELOPMENT.
TEENS, FOR INSTANCE, ARE OFTEN EXAMINED, MORE OFTEN BLAMED, FOR
BEHAVIOR THAT IS OUT OF BOUNDS, FOR MOODINESS, FOR THE SEEMING
IRRESPONSIBILITY THAT MAY COME FROM FOCUS OR JUDGEMENT PROBLEMS.
SOMETIMES ENVIRONMENT, BUT OFTEN GENES ARE BLAMED FOR WHAT CAN
SEEM LIKE MENTAL ILLNESS OR JUST PLAIN 'CRAZY' BEHAVIOR.

WHAT PEOPLE ARE NOT INFORMED ABOUT IS SIMPLE DEVELOPMENTAL PHYSIOLOGY: DIFFERENT REGIONS OF THE TEENAGER'S BRAIN ARE DEVELOPING AT DIFFERENT RATES, AND WHILE MATH SKILLS OR VERBAL SKILLS MAY BE EXTRAORDINARY, THE PART OF THE BRAIN CALLED THE PREFRONTAL CORTEX, WHERE JUDGMENTS ARE FORMED, HAS NOT YET MATURED AND WON'T UNTIL THE TWENTIETH YEAR OR MORE. AND THE LIMBIC SYSTEM, LOCATED DEEP IN THE BRAIN'S INTERIOR WHERE RAW EMOTIONS SUCH AS ANGER ARE GENERATED, IS ENTERING HYPERDRIVE. IT IS POSSIBLE TO SEE THESE REGIONS USING FUNCTIONAL MAGNETIC RESONANCE IMAGING, A TECHNOLOGY THAT TAKES PICTURES OF THE BRAIN'S ACTIVITY. IT TAKES MATURED HARDWARE FOR THE LEARNED SOFTWARE TO FUNCTION PROPERLY.

# SEX, THE CELL, AND MUTATION
## OR: THE EVOLUTION OF INFORMATION CONTENT IN GENES AND BRAINS

HUMANS HAVE 3 BILLION BASE PAIRS OR CHEMICAL LETTERS THAT MAKE UP THE GENOME, INCLUDING THE 100,000 OR SO GENES. A MOUSE HAS 200 MILLION.

HUMANS CARRY MORE GENETIC INFORMATION IN THEIR DNA THAN OTHER MAMMALS (REMEMBERING MORE IS NOT NECESSARILY BETTER THAN) OR AMPHIBIANS OR BACTERIA. IT IS INTERESTING ALSO THAT A LOT OF GENETIC INFORMATION MAY BE REPETITIVE OR FUNCTIONLESS. AND, AS PINKER REMINDS US, "NATURAL SELECTION DOES NOTHING EVEN CLOSE TO STRIVING FOR INTELLIGENCE...THE PROCESS IS DRIVEN BY DIFFERENCES IN THE SURVIVAL AND REPRODUCTION RATES OF REPLICATING ORGANISMS IN A PARTICULAR ENVIRONMENT. OVER TIME THE ORGANISMS ACQUIRE DESIGNS THAT ADAPT THEM FOR SURVIVAL AND REPRODUCTION IN THAT ENVIRONMENT, PERIOD; NOTHING PULLS THEM IN ANY DIRECTION OTHER THAN SUCCESS THERE AND THEN." ONE EXTRA RESOURCE OF HUMANS IS THAT WE HAVE NOT ONLY BIOLOGICAL INFORMATION AVAILABLE TO US, BUT 'EXTRASOMATIC' (OUT-OF-BODY) INFORMATION AS WELL; FOR INSTANCE, SCHOOL, THE LIBRARY, THE INTERNET. LANGUAGE, LIKE THE INTELLECT, MAY BE A SURVIVAL TOOL LIKE ANY OTHER.

"THE RAW MATERIALS OF EVOLUTION ARE MUTATIONS, INHERITABLE CHANGES IN...THE HEREDITARY INSTRUCTIONS IN THE DNA MOLECULE," SAGAN POINTS OUT. MUTATIONS CAN BE CAUSED BY RADIOACTIVITY IN THE ENVIRONMENT; COSMIC RAYS FROM SPACE; OR SIMPLY RANDOMLY BY SPONTANEOUS REARRANGEMENT OF THE NUCLEOTIDES WHICH MUST HAPPEN STATISTICALLY NOW AND THEN. THERE ARE MOLECULES WHICH PATROL THE DNA FOR DAMAGE AND PUT IT TO RIGHT, BUT THIS IS NOT ONE HUNDRED PERCENT EFFICIENT. NOTE: THE PROCESS OF MUTATION IS RANDOM AND STATISTICAL. AFTERWARDS, NATURAL SELECTION KICKS IN, AND SURVIVAL IS AWARDED TO THOSE MOST ADAPTED TO ENVIRONMENT.

AN ACCIDENTAL MUTATION ELONGATED THE NECK OF A GIRAFFE WHO THEN GOT BETTER FOOD AND SO WAS CHOSEN MORE OFTEN AS A MATE AND THEREFORE PRODUCED MORE OFFSPRING THAN THE SHORT-NECKED GIRAFFES. IT WAS NEVER A QUESTION OF SOME GIRAFFE STRETCHING OUT HER NECK FOR BETTER LEAVES AND PASSING ON THIS SPECIALTY. WHAT COUNTS ARE MUTATIONS IN THE EGGS AND SPERM CELLS, WHICH ARE THE AGENTS OF REPRODUCTION, NOT THE EFFORT OF SOME GIRAFFE TO ADAPT BETTER. IF A COSMIC RAY HITS YOU IN THE EYE, IT WILL NOT AFFECT YOUR CHILDREN. IF A COSMIC RAY, OR RANDOM CHANGE IN GENETIC INSTRUCTIONS AFFECTS YOUR EGGS OR SPERM—THIS WILL AFFECT THE PROPAGATION EFFECTS OF YOUR SEX LIFE ON THE SPECIES.

FROM THE PERSPECTIVE OF A GENOME, AN ORGANISM—YOU, FOR INSTANCE—IS JUST A WAY OF COPYING DNA.

WELL, IF THAT ISN'T A DOWNER, I DON'T KNOW WHAT IS.

THINK OF IT THIS WAY. WITHOUT THE WHOLE OF YOU PERSONALLY, YOUR DNA WOULD BE TOTALLY HELPLESS TO COME IN OUT OF THE RAIN.

89

# CHAPTER 7
# INTELLIGENCE AND CONSCIOUSNESS
## in Us, Robots, the Cosmos

## ANIMALS: LEARNING

ALL BUT THE SIMPLEST ANIMALS LEARN AND PROCESS THE INFORMATION THEY LEARN. SOME INFORMATION IS HARDWIRED. SOME IS LEARNED FROM PARENTS AND GROUP BEHAVIOR. ACCIDENTAL, ADAPTIVE, PHYSICAL MUTATION IS THE BASIS FOR CHANGE OF FORM/BEHAVIOR IN ANIMALS. LEARNING ALONE DOES NOT CHANGE THE SPECIES. THE ABILITY TO LEARN IS PASSED ON IN HUMANS—NOT WHAT A PERSON LEARNS IN A LIFETIME.

# HUMANS: LEARNING CAN CHANGE US

## 1 THOUGHT

HUMAN BRAINS HAVE INTELLECT, FRONTAL LOBES THAT CREATE NETWORKS FOR IMAGES, IMAGINATION, INVENTION, TRIAL AND ERROR TESTING, OBSERVATION OF WHAT WORKS, RETENTION AND REFINING OF KNOWLEDGE, OF OBSERVATIONS, INFORMATION.

AND HUMANS CAN PASS ALL THIS ON IN LANGUAGE, WRITING, IDEAS, IN TECHNOLOGY, IN ART, IN RELIGION. LIKE SPIDERS HAVE WEB-SPINNING IN THEIR GENOMES, HUMANS HAVE CULTURE.

*TOO MUCH WORK. I'D RATHER BE A CAT.*

## 2 CONSCIOUSNESS: ALL THE CONTENTS OF OUR BRAIN'S NEURAL NETWORKS

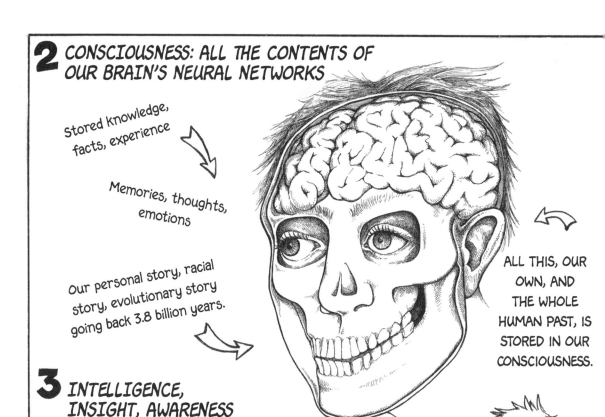

Stored knowledge, facts, experience

Memories, thoughts, emotions

Our personal story, racial story, evolutionary story going back 3.8 billion years.

ALL THIS, OUR OWN, AND THE WHOLE HUMAN PAST, IS STORED IN OUR CONSCIOUSNESS.

## 3 INTELLIGENCE, INSIGHT, AWARENESS

HUMANS HAVE SOMETHING ELSE, SOMETHING MORE—THE CAPACITY FOR NEW INSIGHT INTO OUR OLD THOUGHTS, OUR PAST CONSCIOUSNESS. WE HAVE THE CAPACITY FOR FRESH AWARENESS AND UNDERSTANDING OF THE WORLD INSIDE AND OUTSIDE OUR MINDS. THIS INTELLIGENCE ACTUALLY MUTATES, CHANGES THE WIRING OF OUR BRAINS.

TO SEE SOMETHING DIFFERENTLY IS TO ACT DIFFERENTLY—THIS REWIRES OUR BRAINS AND CHANGES US AND OUR SPECIES THE SAME WAY ANY OTHER PHYSICAL MUTATION CHANGES US.

YOU MEAN IF I CHANGE MY MIND, I CAN CHANGE THE WORLD?

WATCHIT, NARRATOR, YOU'RE GIVING HER A SENSE OF POWER.

YOU BETTER GET IT, CATTIE. ANYONE WHO TRANSFORMS THEIR BRAINS CHANGES THEMSELVES WHICH CHANGES EVERYTHING AROUND THEM. YOU BETTER WATCHIT.

JUST BECAUSE OUR EYES DON'T SEE CORRECTLY, AND WE DON'T EVEN SEE THAT WE DON'T SEE CORRECTLY, WE MISS THE TRUTH. EVERY ATOM AFFECTS EVERY OTHER ATOM. YOU SNEEZE AND THE WHOLE UNIVERSE REACTS! OF COURSE, EVERYTHING YOU DO AND THINK IS IMPORTANT.

# PSYCHOLOGICAL EVOLUTION INTO AWARENESS

PSYCHOLOGICAL AWARENESS IS THE HUMAN BRAIN'S ABILITY TO ACCESS ITS OWN MEMORY BANKS, ITS OWN COMPUTER FILES. WE ARE THE ONLY SPECIES, AS FAR AS WE KNOW, THAT CAN DO THIS. AWARENESS IS A KIND OF PASSWORD.

THE HUMAN BRAIN, BECAUSE IT CAN ACCESS ITSELF, CAN ANALYZE ITS OWN EXPERIENCES, PLAY WITH POSSIBLE COMBINATIONS OF ACTIONS, GENERATE CHOICES.

THE PRICE WE PAY FOR THIS...

WE'RE ALMOST NEVER HERE. HUMAN BRAINS ARE USUALLY EITHER MESSING WITH THE PAST OR HYSTERICAL OVER THE FUTURE. WE DON'T JUST LOOK AROUND AND COMMENT ON THE ACTUAL WORLD AROUND US TO DIRECT OUR ACTIONS OR JUST ENJOY THE SCENERY.

FOR THE FIRST TIME IN EVOLUTION, THE BRAIN CAN ASK ITSELF QUESTIONS, REQUEST ANSWERS TO PROBLEMS. IT CAN REMEMBER THE PAST, PLAN FOR THE FUTURE.

NICE DAY, GO OUTSIDE.

IT'S HOT, MOVE INTO THE SHADE.

LUNCH HAS ARRIVED.

96

# THE MYTH OF THE 'SELF'

THE BIGGEST PRICE WE PAY IS THE DELUSION OF THE 'SELF'. THE HUMAN BRAIN, BECAUSE IT CAN ACCESS ITS OWN PAST STORY, THINKS THERE'S A 'SELF' IN THERE AND WORRIES FEARFULLY ABOUT THAT SELF'S FUTURE SECURITY. OUR ENTIRE CULTURE SUPPORTS THIS MYTH SINCE ALL OUR BRAINS WORK THE SAME WAY.

THE MOST BRILLIANT AMONG US HAVE SEEN THROUGH THIS MYTH OF THE SELF: JESUS, BUDDHA, LAO TZU, KRISHNAMURTI, PINKER, DENNETT, AND MOST BRAIN SCIENTISTS.

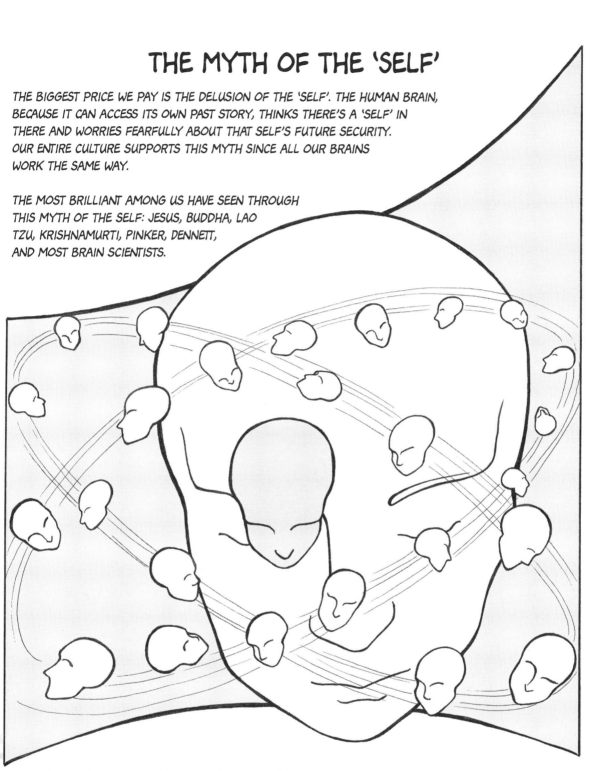

HANS CHRISTIAN ANDERSEN ILLUSTRATED THIS MYTH OF THE SELF IN HIS STORY OF THE LITTLE BOY WHO COULDN'T BE BRAINWASHED INTO SEEING THE EMPEROR'S NEW CLOTHES WHEN HE DIDN'T HAVE ANYTHING ON.

NO MATTER HOW SCIENTISTS SCAN, PROBE, SLICE, PHOTOGRAPH, TEST—THERE'S NO PLACE IN THE BRAIN WHERE THE 'SELF' RESIDES.

## SO—WE'RE NOT SEPARATE FROM. WE'RE ALL ATOMS TOGETHER.

# TEENAGERS & SCIENTISTS

TEENAGERS AND SCIENTISTS HAVE SOMETHING IN COMMON. BOTH HAVE TO MAKE SENSE OF THE WORLD. NEITHER CAN AFFORD TO OBEY BLIND AUTHORITY IN LEARNING THE TRUTH ABOUT ANYTHING.

CLEARLY, TEENS NEED TO PASS EXAMS AND SCIENTISTS DON'T NEED TO REINVENT THE WHEEL OR BRAIN SURGERY. KNOWLEDGE HAS ITS PLACE.

BUT NATURAL SELECTION DID NOT FORM OUR BRAINS TO SWALLOW SECOND-HAND INFORMATION WITHOUT TESTING IT OUT. NATURE FORMED OUR BRAINS TO UNDERSTAND IN ORDER TO SURVIVE IN OUR LOCAL ENVIRONMENT. THE HUMAN BRAIN'S TRILLION SYNAPSES HAS PLENTY OF STORAGE SPACE TO KEEP OUR BODIES AS SAFE AS POSSIBLE FOR OUR BRIEF TWO BILLION SECONDS ON EARTH.

UNDERSTAND: OUR BRAINS HAVE BEEN SELECTED FOR FITNESS, NOT ALWAYS FOR TRUTH. WE SEE WHATEVER IS *OTHER*, WHATEVER IS OUTSIDE OURSELVES, AS MATE OR FOOD, OR AS DANGEROUS.

BUT WE ALSO TRANSLATE *OTHER*, WHATEVER IS DIFFERENT FROM SELF, INTO RACISM, WAR, GREED. AND WE RETAIN IMAGES NOT ONLY FOR SAFETY AND FOOD, BUT WE DISCRIMINATE PSYCHOLOGICALLY IN REMEMBERED ANGER AGAINST ANOTHER

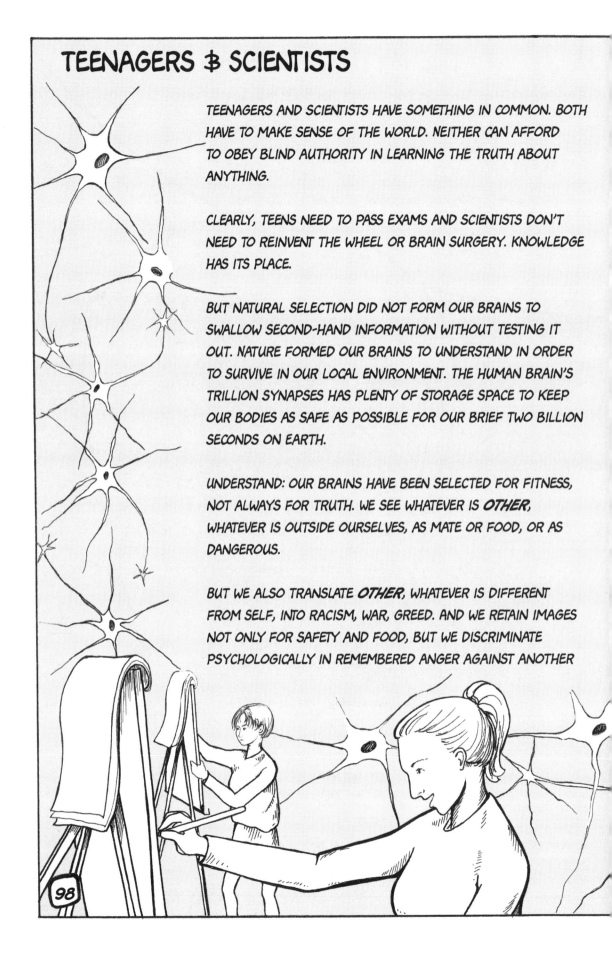

PERSON OR FAMILY OR TRIBE OR NATION OR RELIGION OR
COLOR OR AGE GROUP—AND HERE IS WHERE INTELLIGENCE MUST
INTERVENE WITH THE COGNITIVE FUNCTION, THE COMPUTATIONAL
CATEGORIZATION OF THE BRAIN. THE CATEGORIZATION CANNOT BE
CHANGED DIRECTLY: OUR BRAINS DO TAKE NOTES ON DIFFERENCE.

BUT OUR BEHAVIOR CAN BE ALTERED WITH THIS INTERVENTION OF
INTELLIGENCE OVER INTELLECTUAL CATEGORIZATION.

AWARENESS CAN PRODUCE ETHICS, RIGHT BEHAVIOR. WE CAN
REGISTER A DIFFERENCE WITHOUT HAVING TO OWN IT OR KILL IT.

SCIENTIFIC/RELIGIOUS/PHILOSOPHICAL INQUIRY AND TEEN
BRAINS SHARE PLASTICITY THAT ENCOURAGES INTELLIGENCE AND
THE FREEDOM TO THINK AND BEHAVE DIFFERENTLY FROM THE
PREVIOUS GENERATION AND TO DISCOURAGE COPYCAT BEHAVIOR.
SCIENTISTS AND TEENAGERS ALSO SHARE A DEDICATION TO
CHANGING AND TRANSFORMING THE WAY HUMANS AND THE
WORLD BEHAVES.

LET NO ONE TELL YOU TO GROW UP TOO FAST, SET YOUR BRAIN
IN CEMENT TOO QUICKLY. BE INSECURE. BE UNSETTLED DOWN. BE
UNSET IN YOUR WAYS AS LONG AS POSSIBLE.

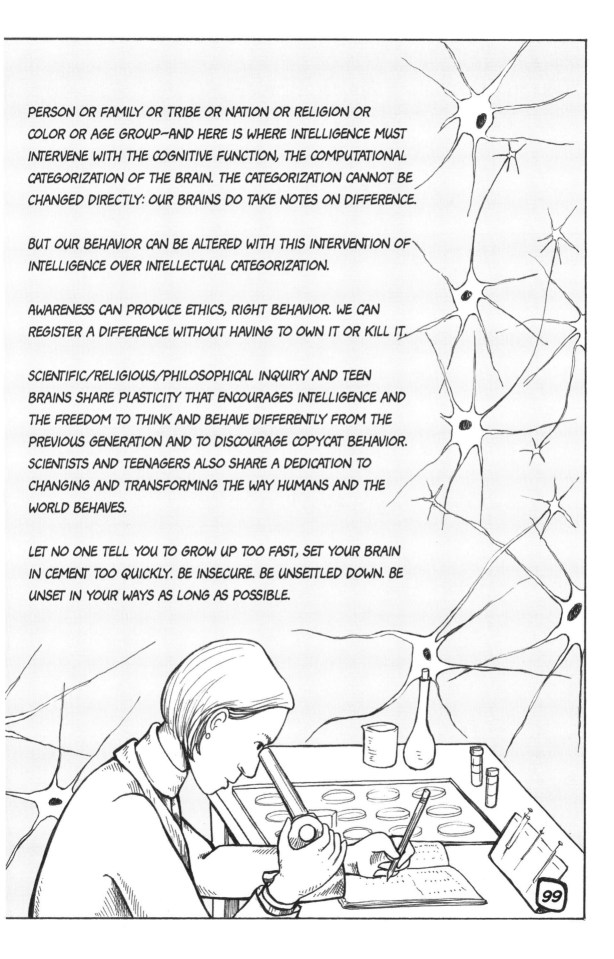

99

# THE POSSIBLE ORIGIN OF CONSCIOUS INTELLIGENCE IS THE COSMOS ITSELF

MY PERSONAL VIEW OF WHERE MIND, INTELLIGENCE, INSIGHT, AWARENESS COMES FROM IS THAT EVERY ATOM, EVERY MOLECULE, EVERY FIELD, EVERY CELL, EVERYTHING IN THE UNIVERSE CONTAINS INTELLIGENCE. IT ISN'T, AS FAR AS MY MENTAL EXPERIMENTS GO, A QUESTION OF FINDING INTELLIGENCE HERE OR THERE OR SOMEWHERE ELSE. INTELLIGENCE IS EVERYWHERE, IN EVERYTHING, AND CONNECTS ALL.

# OTHER THEORIES OF THE HUMAN BRAIN AND ARTIFICIAL INTELLECTS (SURELY NOT INTELLIGENCE): Robots, Satellites, Computers

1. THE HUMAN BRAIN IS SIMPLY A COMPUTER. GENES (INHERITED PHYSICAL BRAIN CAPACITIES) AND MEMES (LEARNED CULTURAL IDEAS) CREATE THE HUMAN BRAIN NETWORK. IT WORKS LIKE ANY ROBOT'S INNARDS OR LAPTOP'S OR SATELLITE'S. WHAT IS PUT IN IS WHAT COMES OUT. A LOT OF HUMANS LIKE THIS THEORY, AS REDUCING LIFE TO MACHINERY GIVES HOPE OF IMMORTALITY.

2. BRAIN SCIENTISTS ARGUE ALL THE TIME ABOUT WHETHER A HUMAN BRAIN CAN BE REPLICATED IN A COMPUTER, WHETHER A ROBOT CAN BE STRUCTURED NOT ONLY TO THINK COGNITIVELY LIKE A HUMAN BEING, BUT TO FEEL, TO HAVE FREE WILL.

3. AMONG THE SECRETS OF LIFE WE HAVE LICKED IS THE CLONING PROCESS. BUT WHEN THEY CLONED DOLLY THE SHEEP, SHE QUICKLY AGED TO DOLLY'S AGE. CLONING YOURSELF WILL NOT ADD YEARS TO YOUR LIFE, IN CASE YOU WERE THINKING ABOUT IMMORTALITY.

4. HUMANS ARE SCARED OF COMING TO AN END, SO WE KEEP TRYING TO THINK OF OURSELVES AS A FORM OF MECHANICAL LIFE WE CAN RECREATE ROBOTICALLY, EVEN BIONICALLY, AND SO LIVE FOREVER.

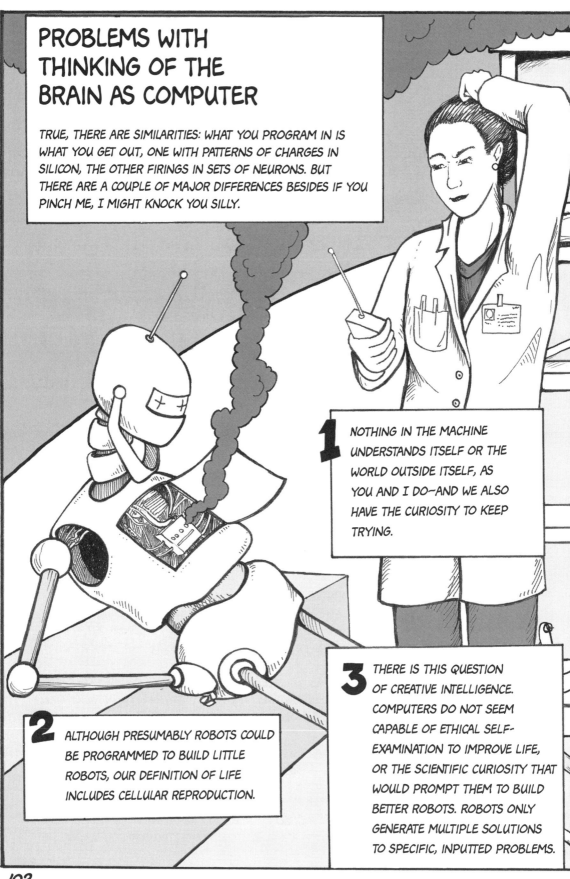

# PROBLEMS WITH THINKING OF THE BRAIN AS COMPUTER

TRUE, THERE ARE SIMILARITIES: WHAT YOU PROGRAM IN IS WHAT YOU GET OUT, ONE WITH PATTERNS OF CHARGES IN SILICON, THE OTHER FIRINGS IN SETS OF NEURONS. BUT THERE ARE A COUPLE OF MAJOR DIFFERENCES BESIDES IF YOU PINCH ME, I MIGHT KNOCK YOU SILLY.

**1** NOTHING IN THE MACHINE UNDERSTANDS ITSELF OR THE WORLD OUTSIDE ITSELF, AS YOU AND I DO—AND WE ALSO HAVE THE CURIOSITY TO KEEP TRYING.

**2** ALTHOUGH PRESUMABLY ROBOTS COULD BE PROGRAMMED TO BUILD LITTLE ROBOTS, OUR DEFINITION OF LIFE INCLUDES CELLULAR REPRODUCTION.

**3** THERE IS THIS QUESTION OF CREATIVE INTELLIGENCE. COMPUTERS DO NOT SEEM CAPABLE OF ETHICAL SELF-EXAMINATION TO IMPROVE LIFE, OR THE SCIENTIFIC CURIOSITY THAT WOULD PROMPT THEM TO BUILD BETTER ROBOTS. ROBOTS ONLY GENERATE MULTIPLE SOLUTIONS TO SPECIFIC, INPUTTED PROBLEMS.

**4** THE AGENTS IN THE MACHINE ARE NONINTELLIGENT, DISCRETE MECHANICAL EXECUTORS OF TASKS. THEY ARE NOT A TINY COPY OF THE ENTIRE SYSTEM REQUIRING THE SYSTEM'S ENTIRE INTELLIGENCE THE WAY A HUMAN BRAIN WORKS. SIMPLY PUT, ALL ROBOTIC BITS ARE REQUIRED TO DO ELECTRONICALLY IS SAY 'YES' OR 'NO' WHEN ASKED.

**5** MOST IMPORTANT, THE COMPUTER CAN ONLY RESPOND FROM MEMORY, FROM PAST INPUT. THE HUMAN BRAIN IS CAPABLE OF INTELLIGENCE, A FREE, UNTETHERED-TO-THE-PAST RESPONSE TO THE PRESENT MOMENT.

GOT IT. HUMANS THINK DIFFERENTLY FROM ROBOTS AND CATS. BUT YOU DON'T TELL ME WHO'S SMARTER!

# THE WAY OF CHANGE AND TRANSFORMATION

HUMAN BRAINS ARE CAPABLE OF FLEXIBILITY, ADAPTABILITY, THE ABILITY TO CHANGE. INSIGHT INTO THOUSANDS OF YEARS OF CONDITIONING IS WHAT TRANSFORMS THE BRAIN.

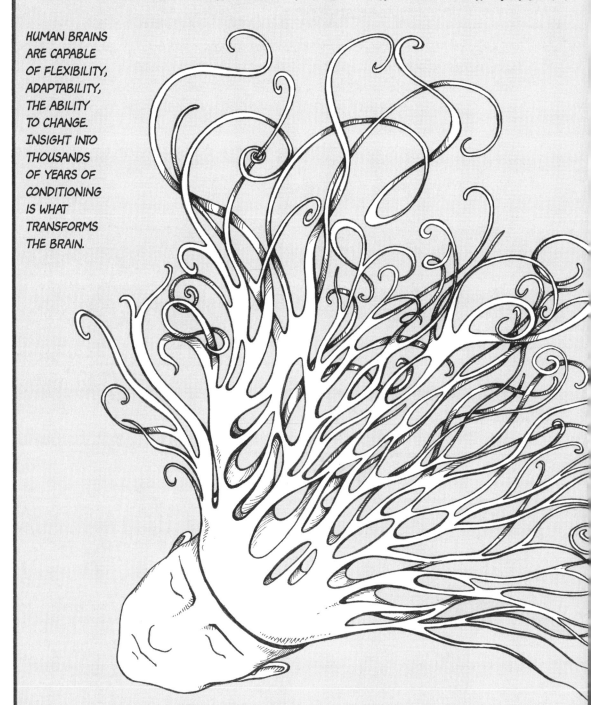

BUT IF COMPLETE TRANSFORMATION OF OUR FEAR-BASED VIOLENCE ISN'T POSSIBLE, AT LEAST THE HUMAN BRAIN CAN, WITH ITS CAPACITY FOR AWARENESS AND INTELLIGENCE, NOT ACT VICIOUSLY ON TERROR WHEN IT STRIKES. WE CAN CHANGE OUR RESPONSES, OUR BEHAVIOR—AND THIS CHANGE WILL ALTER THE NEURONAL STRUCTURE OF THE BRAIN.

WE ARE IN THE MIDDLE OF THE SIXTH EXTINCTION OF SPECIES NOW. AFTER AN EXTINCTION, EVOLUTION GOES INTO OVERDRIVE. THE EARTH AND LIFE WILL GO ON—WITH US OR WITHOUT US.

I PERSONALLY FIND THIS CHEERFUL NEWS.

BUT ONLY IF WE REMAIN IN HARMONY WITH THE PURPOSES OF THE UNIVERSE.

# Glossary of Science Terms

**acceleration:** The rate at which the speed of an object is changing.

**anthropic principle:** We see the universe the way it is because if it were different, we would not be here to observe it.

**antiparticle:** Each type of matter particle has a corresponding antiparticle. When a particle collides with its antiparticle they annihilate each other. This leaves only energy.

**a priori:** Immanuel Kant pointed out that time and 3-D space are not things (phenomena) but only the way human brains are born hard-wired to process what we see. Bohm added, "the subtle mechanism of knowing the truth does not originate in the brain."

**astrophysics:** A branch of astronomy that deals with the physical and chemical constitution of celestial bodies.

**atom:** The basic unit of ordinary matter, consisting of a tiny nucleus (made up of protons and neutrons) and electrons that orbit around the nucleus.

**big bang:** The explosion of a fiery ball that contained all matter and probably even the laws governing it that most cosmologists agree marks the beginning of the universe and time as we know it.

**big crunch:** The contraction of our expanding universe backwards: the end of the universe.

**black hole:** A dense region of space-time from which nothing, not even light, can escape, because gravity is so strong.

**brain:** Brains are, essentially, anticipation machines whose major function is to keep the organism alive in the environment.

**cerebellum:** The brain mass that lies in the back of the head underneath the cerebral cortex.

**chromosomes:** The long strands of hereditary material composed of nucleic acids that contain the genes.

**cognition:** The act or process of knowing based on both intellect and intelligence.

**cortex:** The large outer layer of the cerebral hemispheres, in major part responsible for our characteristically human behaviors.

**cosmology:** The study of the universe as a whole.

**DNA:** Deoxyribonucleic acid is the genetic material of all life on Earth, consisting of ladder-like sequences of units called nucleotides, usually arranged in a double helix. There are two main types of nucleic acids, DNA and RNA, ribonucleic acid.

**entropy:** the degradation of matter and energy in the universe to an ultimate standstill.

**field:** A field exits throughout space and time, as opposed to a particle that exists at only one point at a time.

**four forces:** We (not the universe) have grouped force-carrying particles in four categories according to the strength of the force they carry. Ultimately, physicists hope to find a unified theory that will explain all four forces as different aspects of a single force. The four categories of force are: gravitational force, a universal force of attraction according to mass or energy; electromagnetic force which acts with electrically charged particles but not uncharged particles; weak nuclear force which is responsible for radioactivity; strong nuclear force, which holds the quarks together in the proton and neutron, and the protons and neutrons together in the nucleus of an atom.

**frequency:** For a wave, the number of complete cycles per second.

**frontal lobes:** The frontal lobes of the cortex are involved in judgment, planning, and sequencing of behavior.

**gametes:** Mature sperm or egg cells capable of participating in fertilization. Each gamete has 23 chromosomes, half the number of the 46 in an ordinary body or cell.

**general relativity:** Einstein's theory that the laws of science should be the same for all observers, no matter how they are moving.

**genes:** Genes control the transmission of hereditary characteristics.

**intellect:** Knowledge, information-processing, consciousness. Some centers can be found in the brain through MRI scans.

**intelligence:** Mind, awareness, insight, perception, understanding, the seats of which cannot be found in scanning the physical brain.

**limbic systems:** The structures of the forebrain under the cortex that are concerned with emotion and motivation, such as the hypothalamus, the hippocampus, and the amygdala.

**light year:** The distance traveled by light in one year.

**mass:** The quantity of matter in a body, its resistance to acceleration.

**MRI:** Magnetic resonance imaging is used to scan and observe what parts of the brain are used while a person performs different tasks.

**meme:** A cultural group of ideas that form themselves into distinct units—for example, wheel, jazz, wearing clothes.

**metabolism:** The process by which material is converted to energy for vital activities.

**mind:** Found nowhere physically in the brain, mind, being intelligence itself, may have its origin in the universe itself.

**mitosis:** The process that takes place in a dividing cell.

**mutation:** Inheritable changes in structure and process.

**neutrino:** A particle either so light or so massless it can pass through the Earth and us without hitting anything.

**nuclear fusion:** The process of two nuclei colliding and forming a single, heavier nucleus.

**nucleus:** The central part of an atom, composed only of protons and neutrons.

**phenomenon:** An observable object, fact, event, scientifically describable and known through the senses rather than through thought.

**photon:** smallest possible packet of light energy, a quantum.

**quantum:** Subatomic particles, packet or quantum of light energy.

**quantum theory/quantum mechanics:** The branch of physics that studies the microworld, the behavior of particles and forces in the subatomic world. Much of quantum mechanics is based on Max Planck's theory that light waves were only emitted in packets he called quanta, on the wave/particle duality, and on Heisenberg's Uncertainty Principle—one cannot pinpoint both the position and the velocity of a particle at the same time.

**quark:** A charged elementary particle. Protons and neutrons are each composed of three quarks.

**singularity:** A point in space-time at which the space-time curvature becomes infinite. About fifteen thousand million years ago, the distance between neighboring galaxies was zero, so the density of the universe was infinite as was the curvature of space-time.

**wave/particle duality:** In quantum mechanics, particles may sometimes behave like waves, waves like particles. Otherwise, there is no distinction between particles and waves.

**weight:** The force exerted on a body by a gravitational field.

**zygote:** A fertilized egg.

# Bibliography $ Suggested Reading

Many of the scientists whose work is fundamental to understanding their scientific fields and their relationship to previous, subsequent, and related sciences are mentioned in the text. Some of the books they have written can be understood only by those with science backgrounds, including an understanding of mathematics, the language of physics, or the symbols of chemistry.

The works listed here, however, are accessible to anyone, with or without science background. While they are not written specifically for young adults, they are written for the general audience and very readable, often humorous, and, rather than lecture, provoke thinking on the part of readers. This is particularly true of Stephen Jay Gould, Steven Pinker, and Stephen W. Hawking. I quote also from the philosopher J. Krishnamurti, as he has particularly influenced some of the greatest scientific minds of the twentieth century, especially those of Albert Einstein, Stephen W. Hawking, David Bohm, and Alex Comfort whose work constantly suggests the connection of the human brain with the physical universe.

Major sources for facts and statistics in this book, aside from the following texts, were newspapers, journals, magazines, especially *U.S. News & World Report*, government publications, almanacs, public television specials, documentaries, and broadcasts such as PBS' *NOVA*.

Many of the books listed have already been mentioned in the text of this book.

Carlson, Dale and Hannah Carlson, M.Ed., C.R.C. *Where's Your Head? Psychology for Teenagers*. 2nd edition. Madison, CT: Bick Publishing House, 1998. A general introduction for adults and young adults to the structure of personality formation, the meaning of intelligence, the mind, feelings, behaviors, biological and cultural agenda, and how to transform conditioning in our education systems. Also, *Stop the Pain: Teen Meditations*. Madison, CT: Bick Publishing House, 1999. Self-knowledge is true meditation: ways to lose the anxiety, hurt, conflict, pain, depression, addiction, loneliness, and to move on.

Capra, Fritjof, Ph.D. *The Tao of Physics: An Exploration of the Parallels between Modern Physics and Eastern Mysticism*. 25th anniversary edition. Boston: Shambala, 2000. Philosophical implication of theoretical physics, with an excellent and accessible introduction to relativity theory and atomic physics.

Comfort, Alex, M.D., D.Sc. *Reality and Empathy: Physics, Mind, and Science in the 21st Century*. Albany, State University of New York Press, 1984. An overview, often funny, of the physics, metaphysics, and philosophy of science—physics, biology, mathematics, psychology, and evolutionary theory.

Davies, Paul, Ph.D. *The Fifth Miracle: the Search for the Origin and Meaning of Life*. New York, Simon & Schuster, 1999. Genetic information, space science, physics, chemistry, biology—all are discussed in terms of where life came from, how life began in the first place, is life a chemical fluke, or do we live in a biofriendly universe?

Dawkins, Richard, Phd. *The Selfish Gene*. Oxford, NY: Oxford University Press, 2006. Evolutionary survival of the fittest through natural selection.

Dennett, Daniel C. *Consciousness Explained*. Boston: Little, Brown and Co., 1991. Full-scale exploration of human consciousness, informed by the fields of neuroscience, psychology, and artificial intelligence, this is a funny, clear understanding of the human mind-brain. Not easy reading, but worth it.

Gleick, James, editor. *The Best American Science Writing 2000*. New York: HarperCollins, 2000. Science writers and reporters cover their beat as thoroughly as crime, politics, celebrities are covered. Modern life demands we know about the mapping of DNA, about cloning, about computer science, robots, medical strides and mistakes, about how we got here, about space travel and our place in the universe. A variety of articles culled from the best writers and the best magazines.

Gould, Stephen Jay, Ph.D. *The Mismeasure of Man*. New York: W.W. Norton, 1996. Dr. Gould's challenge to the hereditary I.Q. as a measure of intelligence and destiny. Also *Full House*. New York: Three Rivers Press, 1996. Humans are a twig on the bush of life, not the star on the top of the tree. A funny, funny man, scientist, and writer.

Jordan, Paul. *Early Man*. Gloucestershire, U.K.: Sutton Publishing Ltd., 1999. A brief pocket history covering five million years and the story of human evolution, our collective past.

J. Krishnamurti. *On Education*. London: Krishnamurti Foundation Trust, LTD., 1974. In his searching talks and dialogues with students and teachers, this influential philosopher discusses the differences between intellect and intelligence, brain and mind, right education, and the awakening of a new mind that continues to learn and inquire with both scientific and religious attitudes so that knowledge will not destroy us.

Hawking, Stephen W., Ph.D. *A Brief History of Time: From the Big Bang to Black Holes*. New York: Bantam Books, 1988. A brief, popular, nonmathematical introduction, in words not equations, to astrophysics, the nature and origin of time and the universe. Hawking shows us how our 'world picture' evolved from Aristotle through Galileo and Newton to Einstein and Bohm, and the two great theories of the 20th century: relativity and quantum physics, and the search for the unified theory to resolve all the unsolved mysteries.

Orr, David W. *Ecological Literacy: Education and the Transition to a Postmodern World*. New York: State University of New York Press,1992. The most important discoveries of the 20th century exist not in science, medicine, or technology, but in our dawning awareness of our Earth's limits and how this will affect human evolution. Changes in our educational systems are necessary, and in our lives, to avoid environmental disasters that will match the extinction of the dinosaurs.

Pinker, Steven, Ph.D. *How the Mind Works*. New York: W.W. Norton, 1997. A long but witty, clear, and accessible read by a world expert in cognitive science. Pinker explains what the mind is, how the brain works, how it evolved, how it sees, thinks, feels, enjoys the arts, and ponders the mysteries of life. This is an extraordinary picture of human mental life, with insights that range from evolutionary biology to social psychology.

Sagan, Carl. *The Dragons of Eden: Speculations on the Evolution of Human Intelligence*. New York: Ballantine Publishing Group, 1977. Dr. Sagan once again takes us on a glorious reading adventure that brings together all the sciences that interest human beings, from the big bang to the present, including discussions of death, cloning, computers, intelligent life on other planets, and a guided tour of that lost Eden where dragons ruled. Includes excellent timelines.

Schrodinger, Erwin. *What Is Life? The Physical Aspect of the Living Cell*. Cambridge, U.K.: The Cambridge University Press, 2000. One of the great science classics of the 20th century, this brief book proved to be one of the spurs to the birth of molecular biology and the subsequent discovery of DNA. Questions range from why atoms are so small, to the heredity code and its mutations.

# Science Website Links

Learn more. These websites are a great starting point for your own exploration of science subjects.

**dsc.discovery.com**
Discovery Channel - explore science by subject

**www.google.com/Top/Science/**
Google Science Directory

**www.nationalgeographic.com**
National Geographic, a nonprofit scientific and educational organization

**www.pbs.org/wgbh/nova**
NOVA, a PBS television science series

**www.science.gov**
Science.gov, a gateway to government science information and research results

**www.sciencenews.org**
*Science News*, from the Society for Science and the Public

**www.si.edu**
Smithsonian Institution, the world's largest museum and research complex, dedicated to the increase and diffusion of knowledge

# Teacher's Guide

**SECTION I: WHERE ARE WE FROM?**

**Chapter 1.  Are We the Aliens?**
What is life - Your biological inheritance
Where did life come from - The origin of life

OVERVIEW:
All human beings are the result and contain the record of the entire physical and biological history of the universe, and the evolution of life on earth.

Scientists still don't know where life came from or whether its origin is here on earth or carried on an asteroid from outer space. Important for students to understand that science has not yet found all the answers and that new generations must make their contribution.

OBJECTIVES:
Students will be able to

1.  Understand cosmic history

2.  Explain the difference between life and non-life

3.  Discuss the origin of life

SUMMARY AND EVALUATION:
This is an opportunity for the teacher to sum up class discussion and for the students to pose unanswered questions. For assignment, students may research any of the bio-genesis or physics questions raised.

# Teacher's Guide

**Chapter 2. Your Evolution**
A teenager's cosmic calendar
What is evolution
What is natural selection
What is human evolution

OVERVIEW:
This is a discussion of natural selection of the fittest as the basis of evolution of all species including human evolution as related to the life of the teenager.

OBJECTIVES:
Students will be able to

1. Understand Earth's time scale through human evolution

2. Trace the evolution of species through the human species in relationship to a year in their own lives

3. Discuss how the theory of survival of the fittest works

4. Describe the beginnings of human thought

SUMMARY AND EVALUATION:
Teacher may introduce and assign for classroom debate between Gould's and Pinker's positions: Gould's biological determined potential; Pinker's biological determinism. Or, contrast these points of view with "Does the cosmos have a greater plan?" Or, write an evaluative essay on "How the modern human brain converts chemistry and nerve connections into consciousness" or "How we became the brainiest of all species."

# Teacher's Guide

**Chapter 3. Your Physical Universe**
The visible macro-world - the theory of relativity
The invisible micro-world - quantum theory

OVERVIEW:
How did the universe begin? Is there a prime mover, a great spirit or intelligence in the universe, or was it just always there? Will the universe come to an end and/or will it start up again? What will happen to our Earth? Can we come up with the space technology to get off of Earth to save ourselves when the Sun expands and swallows us? How do the laws of physics get established and if there is a God, did God have a choice in the creation of various parts of the universe, or did God have to obey the various laws of the universe?

OBJECTIVES:
Students will be able to

1.  Understand the structure of the universe, galaxies, and solar systems

2.  Discuss two theories about the origin of all the matter in the universe— the steady state theory and the big bang

3.  Understand All 4 forces of the universe

4.  Understand Classic Physics

5.  Understand Quantum Physics

SUMMARY AND EVALUATION:
Discussion of why the universe exists, what part do we play, and the fate of the universe. We only use 15% of our brain's capabilities. We are still only using thought most of the time, instead of our brain's other capacities; attention to what is happening around us and the second capacity, insight, guides thought to discover answers to these eternal questions. Students brainstorm together to develop a list of 10 behaviors to improve life, living, relating to others, using insight instead of thought.

**Chapter 4. Births of the Universe, Galaxies, Milky Way, and Us**

OVERVIEW:
The big bang gave birth to galaxies, planet systems, and the conditions for life. Provides a detailed timeline.

OBJECTIVE:
Students will be able to

1. Understand what quantum physics does for cosmology and astrophysics, molecular biology is doing for human evolutionary biology. The understanding of the very small, changes our understanding of how larger bodies work.

SUMMARY AND EVALUATION:
Students debate on two schools of thought: 1) life will form under any Earth-like conditions; 2) opposing view, that life on Earth is a unique fluke.

# Teacher's Guide

## SECTION II:  WHERE ARE WE NOW

### Chapter 5.  Your Brain and Its Body
The physical brain
What is consciousness
What is mind
What is electronic thinking
What is mental illness
The body, its parts, and organs

OVERVIEW:
The brain and the body are not two separate units. They are connected. The main job of the brain is to keep the body alive. The physical brain is the primary organ of the nervous system, the control center for all the body's voluntary and involuntary activities. The brain, in its body-mind connection, controls our glands. Love for our young, for each other, may be a physical, inborn, genetic property of animals, based in the limbic system of our bodies, as our adrenalin-based fear and aggression operates from other systems. Question for teacher to students: Does this mean we have an inborn physical capacity to care for and about each other?

OBJECTIVES:
Students will be able to

1.  Understand the physical brain
2.  What is consciousness
3.  What is mind
4.  What is electronic thinking
5.  What is mental illness
6.  The body, its parts, and organs

SUMMARY AND EVALUATION:
Classroom debate: Considering the mind-body, including our glands, could we learn to think differently, behave differently, and not bomb ourselves back into jungles or caves, or fossilize ourselves into dinosaurs?

# Teacher's Guide

**Chapter 6. Genes, Sex, and Mistakes**
DNA - genomics

OVERVIEW:
In his "How the Mind Works", Pinker says, "People don't selfishly spread their genes. Genes spread themselves. They do it by the way they build our brains by making us enjoy life, health, sex, friends, and children. Sexual desire is not people's strategy to propagate themselves. It's people's strategy to attain the pleasures of sex, and the pleasures of sex are the genes strategy to propagate themselves."

OBJECTIVES:
Students will be able to understand

1. Our bodies work pretty well, but because of the purely selfish struggle among living creatures, from the cell to the whole organism, to survive and thus reproduce—the best design is what survives the survival wars.

2. We are the result of previous mistakes (mutations). We are still not perfect.

SUMMARY AND EVALUATION:
Teachers can help students to understand DNA, genomics, and how characteristics get inherited from generation to generation. They can also help students to understand the difference between genes (physical characteristics) and memes (cultural ideas) in the behavior of generation after generation. Have the students list their own personality characteristics into 2 columns and decide whether they are genes or memes.

# Teacher's Guide

**Chapter 7. Intelligence, Consciousness: Us, Robots, and Cosmic**
Is the human brain more than a computer
Is there cosmic intelligence

OVERVIEW:
All but the simplest animals learn by trial and error. Only humans have the capacity for thought which is memory, language, the passing on of information, the preservation in writing and other forms of record keeping. Only humans, as far as we know, have the capacity for intelligence which is to find the truth from the facts we observe.

OBJECTIVES:
Students will be able to

1.  Understand 2 theories about the human brain: 1) it is a computer, and 2) it is more than a computer.

2.  Learn that there is mystery in the universe and decide for themselves whether that Intelligence, Mystery, God, has an actual effect on the universe itself, its physical laws, or each one of us.

SUMMARY AND EVALUATION:
Convey to students that all animals exploit each other in the fight to get ahead. They evolve to cope with the unexpected. We are in the middle of an extinction now. We will be allowed to survive only if we remain in harmony with the purposes of the universe. Ask the students to suggest either in discussion or in essay form how we can attain such a harmony.

# Index

PRESENTS

# A YOUNG ADULT GRAPHIC NOVEL

**COSMIC CALENDAR:**
**From the Big Bang to Your Consciousness**
by Dale Carlson, edited by Kishore Khairnar, Physicist
Illustrations by Nathalie Lewis

Graphic Teen Guide to modern science relates science to teenager's world.

Our minds, our bodies, our world, our universe—where they came from, how they work, and how desperately we need to understand them to make our own decisions about our own lives.

- Teen guide to modern science
- Physics (no equations except for E-mc2)
- Natural history, physical laws of the universe and our Earth
- Evolution and origin of life, DNA and genomics
- Brain and body, intelligence—human, artificial, cosmic
- Dictionary of science terms
- Websites and links for all sciences
- Teacher's Guide and questions

*Cosmic Calendar: Big Bang to Your Consciousness* is the Graphic Edition of the award-winning *In and Out of Your Mind: Teen Science, Human Bites*, a New York Public Library Best Books for Teens.

> "Contemplating the connectivity of the universe, atoms, physics, and other scientific wonders…heady stuff. Carlson delves into the mysteries of Earth, and outer and inner space in an approachable way."
>
> — *School Library Journal*

> "Challenges her readers with questions
> to make them think about the environment, humankind's place in the world, and how ordinary people can help change things for the better."
>
> — *Voice of Youth Advocates*

Illustrations, Index, 160 pages, $14.95,
ISBN: 978-1-884158-34-6

PRESENTS

# LIFE SCIENCE BOOKS FOR TEENS & YOUNG ADULTS

## ARE YOU HUMAN, OR WHAT?
### Evolutionary Psychology for Teens
by Dale Carlson. Pictures by Carol Nicklaus

We have evolved from reptile to mammal to human. Can we mutate, evolve into humane?

- Evolution has equipped us, not for happiness, but for survival and reproduction of the species.

- To survive, we are programmed for fear and pain: every one of us had ancestors who managed to survive, mate, and pass on the best adapted programs for staying alive.

- Our brain programs, hardware and software, have already conquered every other species: we've won, we can stop fighting.

- It's time to pay attention to our psychological welfare as well as our technology.

*"Carlson examines the new science of evolutionary psychology...admirable."*

— *Voice of Youth Advocates*

"Humans have the ability to reprogram their thinking. Humanity will be responsible for its own next psychological step by the choices it makes."

— *School Library Journal*

Illustrations, 224 pages, $14.95,
ISBN: 978-1-884158-33-9

## TALK: TEEN ART OF COMMUNICATION
By Dale Carlson
Foreword by Kishore Khairnar, Dialogue Director

Close, powerful relationships are based on communication: Humans are wired for talk... communication must be learned

Teen Guide to Dialogue and Communication
- How to talk to yourself
- To others
- To parents, teachers, bosses
- To sisters and brothers
- To your best friend
- To groups
- To people you don't like
- To the universe

Author of dozens of books for Young Adults, Carlson has been awarded three ALA Notable Book Awards, the Christopher Award, *ForeWord*'s Bronze Book of the Year, VOYA Honor Book, and is listed on New York Public Library's Best Books for Teens

*School Library Journal* says, "Heady stuff...thought-provoking guide."

The *New York Times Book Review* says of Carlson, "She writes with intelligence, spunk, and wit."

*Publishers Weekly* says, "A practical focus on psychological survival."

"Essential reading."— Jim Cox, *Midwest Book Review*

Illustrations, 192 pages, $14.95,
ISBN: 1-884158-32-3

# LIFE SCIENCE BOOKS FOR TEENS & YOUNG ADULTS

## THE TEEN BRAIN BOOK
### Who & What Are You?
by Dale Carlson. Pictures by Carol Nicklaus.
Edited by Nancy Teasdale, B.S. Physics

Understand your own brain, how it works, how you got the way you are, how to rewire yourself, your personality, what makes you suffer

"Carlson has produced an excellent guide to the teen years." — *ForeWord Magazine*.

**Foreword Bronze Book of the Year, 2004**
Illustrations, Index, 256 pages, $14.95.
ISBN: 1-884158-29-3

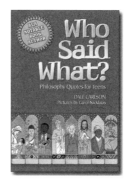

## WHO SAID WHAT?
### Philosophy Quotes for Teens
by Dale Carlson. Pictures by Carol Nicklaus

Teen guide to comparing philosophies of the great thinkers of the ages: form your own philosophy.

"Thought-provoking guide." —School Library Journal

**VOYA Honor Book, 2003,**
**YALSA Quickpicks for Teens 2003**
Illustrations, Index, 256 pages, $14.95.
ISBN: 1-884158-28-5

## IN AND OUT OF YOUR MIND
### Teen Science: Human Bites
By Dale Carlson.
Edited by Kishore Khairnar, M.S. Physics

Teens learn about our minds, our bodies, our Earth, the Universe, the new science—in order to make their own decisions. This book makes science fun and attainable.

"Heady stuff." — School Library Journal

**New York Public Library Best Book for Teens, 2003**
Illustrations, Index, 256 Pages, $14.95
ISBN: 1-884158-27-7

## STOP THE PAIN:
### Teen Meditations
by Dale Carlson. Pictures by Carol Nicklaus

Teens have their own ability for physical and mental meditation to end psychological pain.

• What Is meditation: many ways
• When, where, with whom to meditate
• National directory of resources, centers

"Much good advice...." — *School Library Journal*

**New York Public Library Best Book for Teens 2000**
**Independent Publishers Award**
Illustrations, Index, 224 pages, $14.95;
ISBN: 1-884158-23-4

# LIFE SCIENCE BOOKS FOR TEENS & YOUNG ADULTS

## WHERE'S YOUR HEAD?
### Psychology for Teenagers
by Dale Carlson. Pictures by Carol Nicklaus

- Behaviors, feelings, personality formation
- Parents, peers, drugs, sex, violence, discrimination, addictions, depression, relationship, friends, skills
- Insight, meditation, therapy

"A practical focus on psychological survival skills."
— *Publishers Weekly*

**New York Public Library Books 2000 List**
**YA Christopher Award Book**
Illustrations, Index, 320 pages, $14.95;
ISBN: 1-884158-19-6

## GIRLS ARE EQUAL TOO: The Teenage Girl's How-to-Survive Book
by Dale Carlson. Pictures by Carol Nicklaus

The female in our society: how to change.
- Girls growing up, in school, with boys
- Sex and relationships
- What to do about men, work, marriage, our culture: the fight for survival.

"Clearly documented approach to cultural sexism."
— *School Library Journal*

**ALA Notable Book**
Illustrations, Index, 256 pages, $14.95;
ISBN: 1-884158-18-8

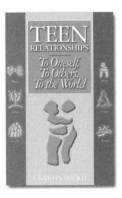

## TEEN RELATIONSHIPS
### To Oneself, To Others, To the World
By J. Krishnamurti. Edited by Dale Carlson

What is relationship? To your friends, family, teachers, in love, sex, marriage, to work, money, government, society, nature, to culture, country, the world, God, the universe

J. Krishnamurti spoke to young people all over the world. "When one is young." he said, "one must be revolutionary, not merely in revolt… to be psychologically revolutionary means non-acceptance of any pattern."

Illustrations, Index, 288 Pages, $14.95.
ISBN: 1-888004-25-8

## WHAT ARE YOU DOING WITH YOUR LIFE?
### Books on Living for Teenagers
By J. Krishnamurti. Edited by Dale Carlson

Teens learn to understand the self, the purpose of life, work, education, relationships.

The Dalai Lama calls Krishnamurti "one of the greatest thinkers of the age." Time magazine named Krishnamurti, along with Mother Teresa, "one of the five saints of the 20th century."

**Translated into five languages.**
Illustrations, Index, 288 Pages, $14.95.
ISBN: 1-888004-24-X

PRESENTS

# FICTION FOR TEENS & YOUNG ADULTS

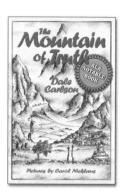

## THE MOUNTAIN OF TRUTH
Young Adult Science Fiction
by Dale Carlson
Illustrated by Carol Nicklaus

Teenagers sent to an international summer camp form a secret order to learn disciplines of mind and body so teens can change the world.

"Drugs and sex, both heterosexual and homosexual, are realistically treated…incorporates Eastern philosophy and parapsychology."

— *School Library Journal*

**ALA Notable Book**
**ALA Best Books for Young Adults**
224 pages, $14.95, ISBN: 1-884158-30-7

## THE HUMAN APES
Young Adult Science Fiction
by Dale Carlson
Illustrated by Carol Nicklaus

Teenagers on an African expedition to study gorillas meet a hidden group of human apes and are invited to give up being human and join their society.

"A stimulating and entertaining story."

— *Publishers Weekly*

**ALA Notable Book**
**Junior Literary Guild Selection**
224 pages, $14.95, ISBN: 1-884158-31-5

PRESENTS

## BOOKS FOR HEALTH & RECOVERY

**THE COURAGE TO LEAD—Start Your Own Support Group: Mental Illnesses & Addictions**
By Hannah Carlson,
M.Ed., C.R.C.

Diagnoses, Treatments, Causes of Mental Disorders, Screening tests, Life Stories, Bibliography, National and Local Resources.

"Invaluable supplement to therapy."

— *Midwest Book Review*

Illustrations, Index, 192 pages, $14.95; ISBN: 1-884158-25-0

**CONFESSIONS OF A BRAIN-IMPAIRED WRITER**
A Memoir by Dale Carlson

"Dale Carlson captures with ferocity the dilemmas experienced by people who have learning disabilities… she exposes the most intimate details of her life….Her gift with words demonstrates how people with social disabilities compensate for struggles with relationships."

— Dr. Kathleen C. Laundy, Psy.D., M.S.W., Yale School of Medicine

224 pages, $14.95, ISBN: 1-884158-24-2

**STOP THE PAIN:**

**Adult Meditations**
By Dale Carlson

Discover meditation: you are your own best teacher. How to use meditation to end psychological suffering, depression, anger, past and present hurts, anxiety, loneliness, the daily problems with sex and marriage, relationships, work and money.

"Carlson has drawn together the diverse elements of the mind, the psyche, and the spirit of science…Carlson demystifies meditation using the mirrors of insight and science to reflect what is illusive and beyond words."

— R.E. Mark Lee, Director, Krishnamurti Publications America

Illustrations, 288 pages, $14.95; ISBN: 1-884158-21-8

## PRESENTS

| **BOOKS ON LIVING WITH DISABILITIES** | **BOOKS ON WILDLIFE REHABILITATION** |

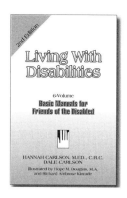

**LIVING WITH DISABILITIES**
**6-Volume Compendium**
by Hannah Carlson, M.Ed., CRC • Dale Carlson

Endorsed by Doctors, Rehabilitation Centers, Therapists, and Providers of Services for People with Disabilities

INCLUDES:
*I Have A Friend Who Is Blind*
*I Have A Friend Who Is Deaf*
*I Have A Friend With Learning Disabilities*
*I Have a Friend with Mental Illness*
*I Have A Friend With Mental Retardation*
*I Have A Friend In A Wheelchair*

*"Excellent, informative, accurate."*

> — Alan Ecker, M.D., Clinical
> Associate Professor at Yale

ISBN: 1-884158-15-3, $59.70

**WILDLIFE CARE FOR BIRDS AND MAMMALS**
**7-Volume Compendium**
by Dale Carlson and Irene Ruth

Step-by-Step Guides • Illustrated
Quick Reference for Wildlife Care
For parents, teachers, librarians who want
to learn and teach basic rehabilitation

INCLUDES:
*I Found A Baby Bird, What Do I Do?*
*I Found A Baby Duck, What Do I Do?*
*I Found A Baby Opossum, What Do I Do?*
*I Found A Baby Rabbit, What Do I Do?*
*I Found A Baby Raccoon, What Do I Do?*
*I Found A Baby Squirrel, What Do I Do?*
*First Aid For Wildlife*

ISBN: 1-884158-16-1, $59.70

**First Aid For Wildlife**
ISBN: 1-884158-14-5, $9.95
Also available separately.

*Endorsed by Veterinarians,
Wildlife Rehabilitation
Centers, and National
Wildlife Magazines.*

# ORDER FORM

307 NECK ROAD, MADISON, CT 06443
TEL. 203-245-0073 • FAX 203-245-5990
www.bickpubhouse.com

Name: _____

Address: _____

City: _____ State: _____ Zip: _____

Phone: _____ Fax: _____

| QTY | BOOK TITLE | PRICE | TOTAL |
|---|---|---|---|
| | **YOUNG ADULT FICTION** | | |
| | The Human Apes | 14.95 | |
| | The Mountain of Truth | 14.95 | |
| | **YOUNG ADULTS/GRAPHIC NOVEL** | | |
| | Cosmic Calendar: From the Big Bang to Your Consciousness | 14.95 | |
| | **YOUNG ADULTS/TEENAGERS NONFICTION** | | |
| | Are You Human or What? | 14.95 | |
| | Talk: Teen Art of Communication | 14.95 | |
| | The Teen Brain Book | 14.95 | |
| | Who Said What? Philosophy Quotes for Teens | 14.95 | |
| | In and Out of Your Mind: Teen Science: Human Bites | 14.95 | |
| | What Are You Doing with Your Life? | 14.95 | |
| | Stop the Pain: Teen Meditations | 14.95 | |
| | Where's Your Head?: Psychology for Teenagers | 14.95 | |
| | Girls Are Equal Too: The Teenage Girl's How-To-Survive Book | 14.95 | |
| | **BOOKS FOR HEALTH & RECOVERY** | | |
| | The Courage to Lead | 14.95 | |
| | Confessions of a Brain-Impaired Writer | 14.95 | |
| | Stop the Pain: Adult Meditations | 14.95 | |
| | **BOOKS ON LIVING WITH DISABILITIES** | | |
| | Living with Disabilities | 59.70 | |
| | **BOOKS ON WILDLIFE REHABILITATION** | | |
| | Wildlife Care for Birds and Mammals | 59.70 | |
| | First Aid for Wildlife | 9.95 | |
| | TOTAL | | |
| | SHIPPING & HANDLING: $4.00 (1 Book), $6.00 (2), $8.00 (3-10) | | |
| | **AMOUNT ENCLOSED** | | |

Send check or money order to Bick Publishing House. Include shipping and handling.

**Also Available at your local bookstore from: BookWorld,
Baker & Taylor Book Company, and Ingram Book Company**

## AUTHOR

**Dale Carlson**

Author of over fifty books, adult and juvenile, fiction and nonfiction, Carlson has received three ALA Notable Book Awards, and the Christopher Award. She writes science, philosophy, and psychology books and science fiction novels for young adults, and general adult nonfiction. Among her titles are *The Mountain of Truth* (ALA Notable Book), *Girls Are Equal Too* ( ALA Notable Book), *Where's Your Head?: Psychology for Teenagers* (Christopher Award, N.Y. Public Library Best Books List), *Stop the Pain: Teen Meditations* (N.Y. Public Library Best Books List), *Who Said What? Philosophy for Teens* (VOYA Honor Book), *The Teen Brain Book: Who and What Are You?* (ForeWord Magazine Book of the Year), *Talk: Teen Communication* (ForeWord Magazine Finalist), *Wildlife Care for Birds and Mammals*. *Stop the Pain: Adult Meditations* follows her teen meditation book. Carlson has lived and taught in the Far East: India, Indonesia, China, Japan. She teaches writing here and abroad during part of each year.

## ILLUSTRATOR

**Nathalie Lewis**

Nathalie Lewis first earned a B.F.A. in Fine Arts from Laval University in Quebec, Canada. She then went on to pursue advanced studies in Anthropology and worked in Archeology for the city of Quebec for several years. She is now a free lance illustrator. She lives in New England.

# EDITOR

**Kishore Khairnar, M.Sc. Physics**

Gold-medal physicist, professor of physics, mathematics, and computer science, Kishore Khairnar worked with ITT, and the Electronic & Engineering Company in Mumbai, before creating his own NDT (Non-Destructive Testing) Company to provide engineering inspection services to industries all over India. He then joined the Krishnamurti Foundation India. He taught sixth to twelfth grade physics, electronics, and mathematics at the Rajghat Education Centre, organized dialogues and international gatherings for the study of the great philosopher Krishnamurti's teachings. As Chief Archivist at Rajghat in India and Brockwood in England, he worked at creating and computerizing the complete published works of Krishnamurti. He is publisher of the first Indian language translation of Krishnamurti's life, and of several English collections of the teachings in India. He is currently Director of the Krishnamurti Study Centre at Sahyadri, Pune, India. Khairnar is also coeditor with Dale Carlson of the American edition of *What Are You Doing With Your Life?*, the first Krishnamurti collection of teachings for teenagers. He and his wife, poet and music critic Professor Kalyani Inamder, live in Pune at the Sahyadri Krishnamurti Study Centre where he continues to publish, edit, and conduct dialogues.